PANELING
WITH
SOLID LUMBER
INCLUDING PROJECTS

"For since the creation of the world God's invisible qualities—his eternal power and divine nature—have been clearly seen, being understood from what has been made, so that men are without excuse."

(Romans 1:20).

"He has showed you, O man, what is good. And what does the Lord require of you? To act justly and to love mercy and to walk humbly with your God."

(Micah 6:8 NIV).

Other TAB Books by the Author

No. 1868
$18.95

PANELING
WITH
SOLID LUMBER
INCLUDING PROJECTS

DAN RAMSEY

TAB BOOKS Inc.
Blue Ridge Summit, PA 17214

FIRST EDITION

SECOND PRINTING

Printed in the United States of America

Reproduction or publication of the content in any manner, without express
permission of the publisher, is prohibited. No liability is assumed with respect to
the use of the information herein.

Copyright © 1985 by TAB BOOKS Inc.

Library of Congress Cataloging in Publication Data

Ramsey, Dan, 1945-
Paneling with solid lumber, including projects.

Includes index.
1. Paneling—Amateurs' manuals. 2. Lumber—Amateurs'
manuals. I. Title.
TH8581.R36 1985   643'.7   85-4652
ISBN 0-8306-0868-0
ISBN 0-3806-1868-6 (pbk.)

Questions regarding the content of this book
should be addressed to:

Reader Inquiry Branch
Editorial Department
TAB BOOKS Inc.
Blue Ridge Summit, PA 17294

Front cover photographs: Top left photograph courtesy of Georgia Pacific Cor-
poration. Remaining photographs were taken by the author.

Contents

PANELING
WITH
SOLID LUMBER
INCLUDING PROJECTS

Acknowledgments

A CREATIVE WRITER IS ONE WHO GATHERS from the greatest number of sources. As the writer of this book on solid lumber paneling projects, I want to acknowledge the many sources that have offered information and illustrations to make it as complete as possible.

In random order, these sources include P. Allsebrook of California Redwood Association; Susan Soorts Williams of Georgia-Pacific Corp.; Martha Nold of Champion International; Clark E. McDonald of Hardwood Plywood Manufacturers Association; Richard Wallace of Southern Forest Products Association; Leo Floros of Selz, Seabolt & Associates for Aromatic Red Cedar Closet Lining Manufacturers Association; Dan Herman of Simpson Strong-Tie Co., Inc.; Potlatch Corp.; Washington Energy Extension Service; Bureau of Naval Personnel; Department of the Army; and the U.S. Department of Agriculture.

Introduction

THE NEWEST IDEA IN DO-IT-YOURSELF INTER-
ior decorating is solid lumber paneling. Solid
lumber paneling is interior paneling made of solid
pieces of lumber, rather than plywood, and installed
on walls, ceilings, and cabinets. Solid lumber paneling is offered in widths from 2 to 12 inches and standard or random lengths of solid pine, fir, cedar,
redwood, and other woods that can be easily installed in many decorative patterns. A few simple
hand tools and this book are all you need to turn
an ordinary wall into a warm and rich addition to
any room with solid lumber paneling.

Paneling with Solid Lumber, including Projects
offers simple step-by-step instructions on every aspect of planning, selecting, preparing, installing,
fastening, and finishing solid lumber panels. It also
offers numerous projects, both easy and fancy, that
you can make with solid lumber paneling.

The illustrations in this book will make it clear
that you can add beauty and function to your home,
business, or summer house with easy-to-install solid
lumber paneling projects.

Chapter 1

Planning Paneling Projects

EVERY WALL IN YOUR HOME IS LIKE A BLANK canvas, and you have many options in decorating each wall. Ordinary paint, wall covering, or even carpeting can be used, but more homeowners are selecting solid lumber paneling. It adds warmth and style when combined with painted, mirrored, or tiled walls. It provides an exciting counterpoint to walls covered with wall covering or fabric. Solid lumber paneling adds a vibrant, living quality to your home (Figs. 1-1 through 1-3).

The beauty of solid lumber paneling is more than just what meets the eye. It's also a matter of dollars and cents. Wood paneling wears better than other wall treatments. It adapts to any refurnishing; hence it never goes out of fashion. Solid lumber paneling adds a richness to walls and other interior surfaces that is hard to equal with almost any other building or decorating material.

Decorators consider wood a natural choice for many schemes and paneling a natural coordinator for any decor. It enhances wood furniture tones and

is the ideal background for upholstery and drapery fabrics. It works beautifully in combination with metals, plastics, and glass. Just about everything you have or plan for your home will look its best against a background of solid lumber paneling. Best of all, solid lumber paneling is simple and quick to install by even the first-time do-it-yourselfer.

CHOOSING SOLID LUMBER PANELING

Generally, the smoother wood grains with the look of fine furniture finishes are the best paneling choices for more formal rooms. Consider them for living rooms, master bedrooms, dining areas, or foyers. Panel the entire space or combine these types of solid lumber paneling with formal wall coverings or traditionally formal paint tones (Figs. 1-4 through 1-11).

Paneling with rustic character, such as those with pronounced wood grain texture and knots, or weathered finishes is the natural choice for casual areas. Use them for family and children's rooms, dining nooks, and country kitchen effects. Good

1

Fig. 1-1. A typical basement room before the installation of solid lumber paneling (courtesy California Redwood Association).

Fig. 1-2. Same basement wall with solid lumber paneling (courtesy California Redwood Association).

Fig. 1-3. Solid lumber paneling adds a wood texture to interior decorating (courtesy California Redwood Association).

Fig. 1-4. Solid lumber panel graining adds warmth to any room (courtesy Hardwood Plywood Manufacturers Association).

coordinates include country-styled or contemporary wall coverings and almost all paint tones.

Like most decorating suggestions, however, these are not hard and fast rules. Solid lumber paneling is so versatile it can add to any scheme, no matter what your preference.

Paneling the entire room with solid lumber paneling is one option. Combining it with another wall treatment is also an attractive option. Study each room to see what direction suits the room best. Totally paneled rooms have a wonderful finished look. For best effects, match the paneling to such considerations as how bright the room is in daylight and at night, and whether you want the room to look cozier or to seem more spacious. To make a large room or a room with cold northern exposure seem cozy and cordial, consider medium tones of

Fig. 1-5. Diagonal solid lumber paneling in an entryway (courtesy California Redwood Association).

solid lumber paneling and finish or even dark, warm tones. To make a small room or a room with bright southern exposure seem cool and spacious, consider the light to medium tones of solid lumber paneling and those with weathered, grey finishes. To make a room appear longer, panel the far end to draw the eye.

Altering the grain direction gives an entirely different look to paneling. Consider using paneling horizontally, on the diagonal, or even in a chevron pattern to point up one area or other.

Panel one entire wall to give a room an elegant focus, use solid lumber paneling to set off one section, such as the dining area in an open plan. Extend touches of paneling to adjacent areas by using wainscotting for a handsomely coordinated wall motif.

PANELING PLANS

Once you've decided to install solid lumber paneling in your home or office, you'll want to discover how much is needed before you set your budget and purchase the paneling. A professional-looking and trouble-free project always starts with essential planning and preparation (Fig. 1-12). Before you install the first paneling board, you need to calculate the paneling required.

Begin by measuring the room or rooms and transferring those measurements to graph paper, as shown in Figs. 1-13 and 1-14. A graph diagram

4

Fig. 1-6. Solid lumber paneling decorates a living room (courtesy California Redwood Association).

Fig. 1-7. Solid lumber paneling can add beauty to a den or family room (courtesy California Redwood Association).

Fig. 1-8. This bathroom's wall and ceiling are decorated with solid lumber paneling (courtesy California Redwood Association).

Fig. 1-9. Warm outdoor tones come from diagonal solid lumber paneling in the bath area (courtesy California Redwood Association).

Fig. 1-10. Sunlight enhances the texture of solid lumber paneling in this spa (courtesy California Redwood Association).

Fig. 1-11. An attic can become a beautiful room with the addition of solid lumber ceiling paneling (courtesy California Redwood Association).

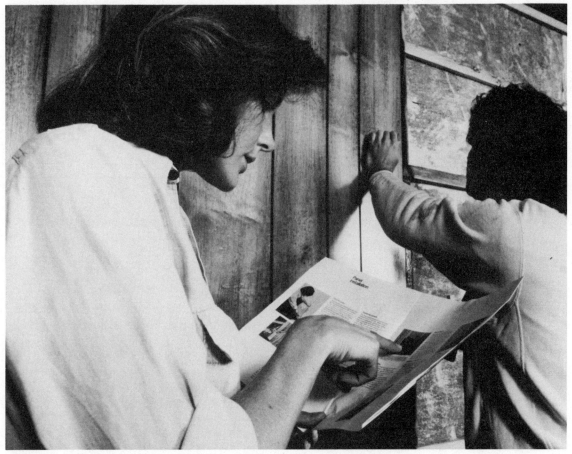

Fig. 1-12. The first step to installing paneling is planning (courtesy Georgia-Pacific).

also helps you visualize the project, including furnishings. If your room is 14 × 16 feet and each 1/4-inch square equals 6 inches, you'll end up with a 7-×-8-inch rectangle. Indicate windows, doors, and other structural elements such as fireplaces.

Computing the amount of lumber needed for baseboards, moldings, casings, and trim requires adding up the total linear feet each will cover. You may want to mark these needs on your plan with a colored pen (Fig. 1-15).

To figure the amount of lumber needed for paneling, determine the square footage of your walls. Convert this figure to board and linear feet of lumber using the simple conversion factors provided in Table 1-1.

First, multiply each wall's total width by its

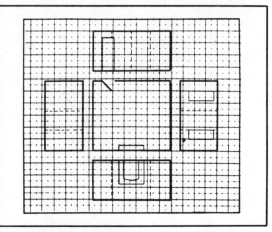

Fig. 1-13. Laying out a room on graph paper (courtesy Georgia-Pacific).

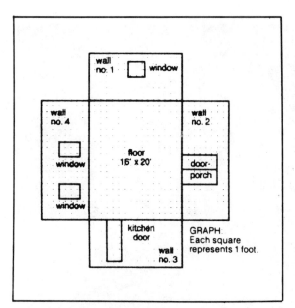

Fig. 1-14. Specific location of doors and windows will help in planning solid lumber paneling (courtesy Champion International).

Table 1-1. Wall Square
Foot and Linear Foot Conversions.

LUMBER WIDTH	SQUARE FOOT CONVERSION	LINEAR FOOT CONVERSION	⅜ INCH PANELING ONLY
4"	1.24	3	1.19
6"	1.15	2	1.12
8"	1.11	1½	1.09
10"	1.09	1-1/5	(wider lumber
12"	1.07	1	not available)

height and subtract nonpaneled areas, windows, and doors. Then multiply this figure by the board and linear foot conversion factors appropriate for the width of the lumber you select. Paneling 100 square feet of wall space with 6-inch-wide tongue-and-groove lumber requires 115 square feet (100 × 1.15 = 115) or 230 linear feet (115 × 2 = 230). Be sure to allow 5 percent extra for errors and end trim loss, and 15 percent extra for diagonal paneling. Table 1-2 offers coverage estimates for walls and other surfaces with solid lumber paneling.

Chapter 2 will guide you in selecting specific

Fig. 1-15. Colored pens will help you mark your plan for molding, casing, and trim requirements (courtesy Georgia-Pacific).

solid lumber paneling woods, grades and joints. To help you to better plan your decorating project, let's first consider the many elements involved in interior decoration and color harmony.

COLOR HARMONY

Color is the art in decorating. It is the most difficult part of decorating to the inexperienced. It requires more careful concentration than any of the other skills that decorating develops. You can install solid lumber paneling and other decorating elements with limited experience. The selection of colors and tints, however, requires experience and knowledge of the factors involved (Fig. 1-16).

If you understand what color is, many of the problems simplify themselves (Figs. 1-17 through 1-20). To begin with, pigments in wood, stains, and paint do not in themselves have any color, but each pigment does reflect light of a definite wavelength. When seen by the retina of the eye and interpreted by the brain, a particular lightwave causes the sensation of a particular color. Without light or no color could be seen. Because of these circumstances, the light source is as important as the colors used.

Problems of color used on the exterior of a house are simplified by the fact that the light source is not variable. Interior painting, on the other hand, involves variations among sunlight, incandescent light, and fluorescent light, as well as the color effects of lamp shades, all of which must be kept in mind when planning a color scheme. The color surfaces of your home furnishings, flat wall colors, and even skin color take on a startling variety of

12

Paneling			
Dimensions in Inches			*Quantity Reqd. to Cover 1,000 Sq. Ft. of Wall Area (in Surface Measure)
Nominal Size	Thickness	Width	
1 × 4	3/4	3 1/2	1231
1 × 5	3/4	4 1/2	1177
1 × 6	3/4	5 1/2	1143
1 × 8	3/4	7 1/4	1104
1 × 10	3/4	9 1/4	1084
1/2 × 4	7/16	3 1/2	1231
1/2 × 5	7/16	4 1/2	1177
1/2 × 6	7/16	5 1/2	1143
1/2 × 8	7/16	7 1/4	1104
*Allow Small Additional Footage for Cutting and Fitting.			

Table 1-2. Paneling Coverage.

shadings depending on the type of lighting in your home.

There are six, perhaps more, types of illuminants to consider when choosing color harmony. These are: natural daylight, incandescent filament lamps, and at least four basic kinds of fluorescent lamps with more types being developed all the time (see my book *Effective Lighting for Home and Business* (TAB Book No. 1658)). You should observe, when purchasing lighting, what effects will result in the colors and textures used in your interior decorating, especially your wood surfaces.

THE COLOR WHEEL

The color wheel is based upon the fact that with three hues—red, yellow and blue—you can mix any other color you wish. These three colors are called the *primaries* (Fig. 1-21). Yellow and red when mixed in equal amounts will give you a good orange; yellow and blue will give you green; blue and red will give you purple. These three mixed colors are called the *secondaries* (Fig. 1-22).

Intermediate colors between the primaries and secondaries should be mixed next to complete a

12-color wheel (Fig. 1-23). With a color wheel you can learn a great deal about combining colors harmoniously.

Color harmony means simply the congruity of hue, intensity, value, and contrast. You must choose colors so that they look right together, and they will look right if you follow the rules for color harmony.

Hue is a particular color; that which distinguishes the color from any other color. *Intensity* is the purity of the color or hue. The closer it is to the primary, the more intense it is. *Value* indicates the lightness or darkness of a hue. If it is very light from the addition of white pigment, it is said to have high value. If it is very dark from the addition of black, it is said to have a low value. *Contrast* is the use of different colors, values, or intensities to achieve color harmony.

RULES OF COLOR HARMONY

The rules of color harmony can be reduced to two, with related variations. The simplest rule is that of *analogous harmony*. If you look at the 12-color wheel and block out with a sheet of white paper all but three adjacent hues, you will find that the combination of any two or three is pleasant. Experimenta-

Fig. 1-16. Solid lumber paneling should be selected to match both furniture and personal tastes (courtesy Champion International).

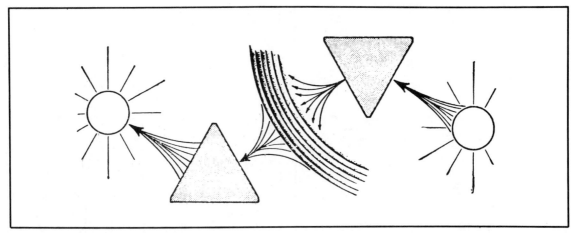

Fig. 1-17. Clear sunlight contains all the colors, which may be demonstrated by passing it through a prism. The result is a rainbow of hues.

tion will prove that the value of any of these colors can be changed at will with results that may be even more pleasing. Keeping the value of one of the three at its original intensity and changing the values of the other two will introduce you to the decorator's favorite trick of having an accent color in your scheme.

You now have the first basic rule: *Colors adjacent to one another on the color wheel are in harmony.* Variations are achieved by the use of different

values, which will also provide you with variations in intensity.

The second rule is that of *complementary harmony*. Refer to Fig. 1-24. On the color wheel you will find that green is opposite red. For some persons, this makes a happy combination. Just think of Christmas decorations. Eastertime decorations illustrate equally well the congruity of yellow and purple. While you don't want to paint a room half red and half green, you will find that a gray-green

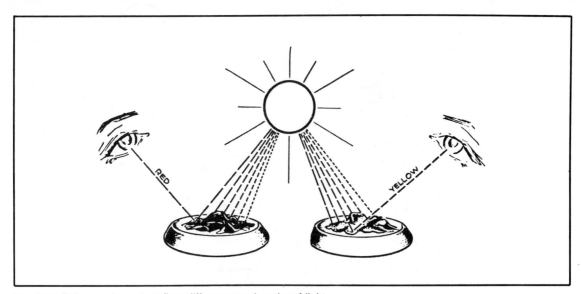

Fig. 1-18. Different pigments reflect different wavelengths of light.

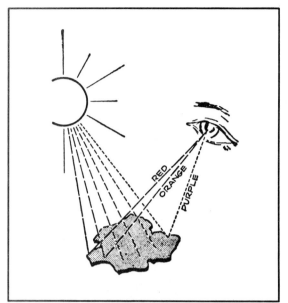

Fig. 1-19. Pigments don't reflect just one wavelength of light. This is one cause of apparent color changes.

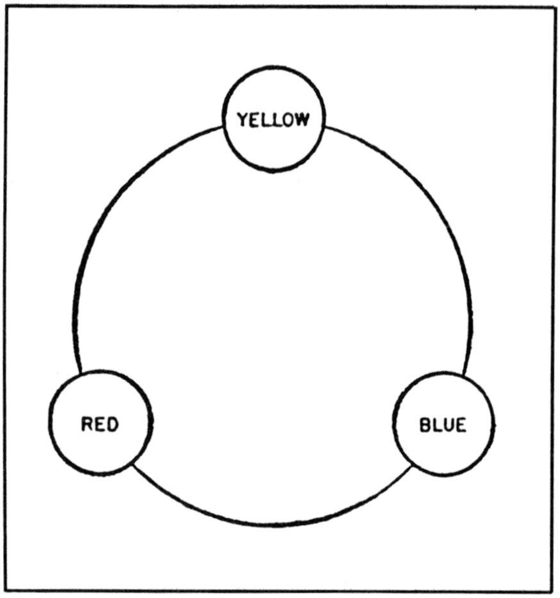

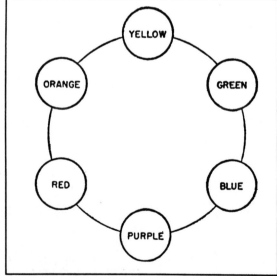

Fig. 1-21. Primary colors: red, yellow, blue.

room is complemented with red cedar solid lumber paneling. The second basic rule of color harmony then, is: *Colors opposite to one another on the color wheel are in harmony.*

You can vary the values or intensities of complementary hues. In addition, you can apply the rules of analogous harmony to complementary harmony; you can use either or both of the adjacent

Fig. 1-20. Even a "daylight" incandescent light is considerably yellower than sunlight.

Fig. 1-22. Secondary colors.

16

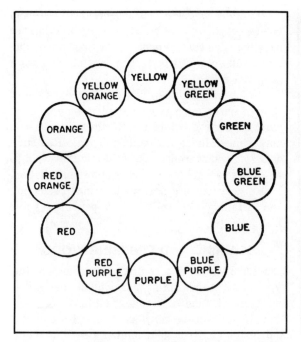

Fig. 1-23. A 12-color color wheel shows the primaries, the secondaries, and the first range of intermediates.

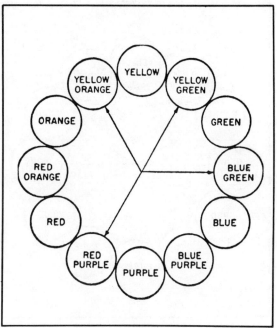

Fig. 1-25. A variation of split complementary harmony.

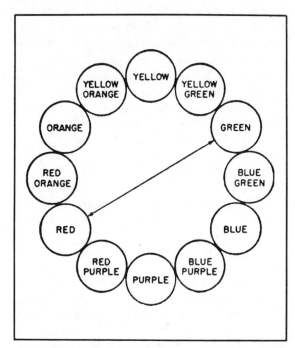

Fig. 1-24. Complementary harmony: colors opposite to one another are in harmony.

hues of one of the original colors to form a harmony known as *split complementary*. Refer to Fig. 1-25.

When you are familiar with basic harmony, try more subtle ones based upon variations of complementary harmonies. There are *double complements* (Fig. 1-26). For instance, yellow-orange with blue-purple might be one pair and yellow-green with red-purple the other. In such combinations, however, choose carefully.

A third possibility is *triad harmony* (Fig. 1-27). This is not a true complement, but if great caution is used in handling the intensities, the results can be very pleasing. Triad harmony is simply the selection of any three hues that are equally spaced around the color wheel, like red, yellow, and blue.

USING COLOR IN PANELING

So what does all this mean to the person wanting to install solid lumber paneling? Many things. Since solid lumber paneling adds a great mass of color to a given room, you will want to know how this color will affect the room. Will it accent or detract?

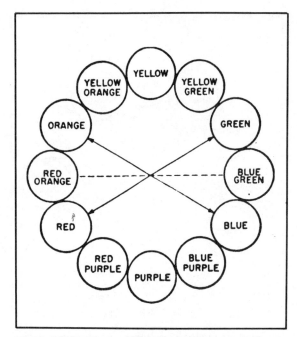

Fig. 1-26. Double complementary harmony: a subtle scheme calling for caution and imagination in the variation of intensities.

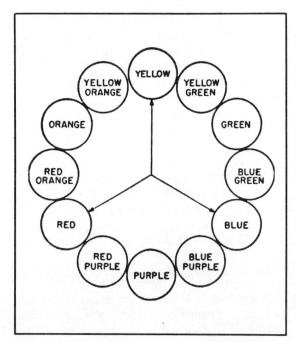

Fig. 1-27. Triad harmony. The colors are equally spaced around the color wheel.

In addition, color consciousness will also help you select the type of wood for your solid lumber paneling. The current decor of your room may dictate a soft color of wood, such as white pine, or a deep red cedar.

Color will also help you decide what type of stain or finish you may wish to place on your solid lumber paneling before or after installation. You may want to finish with a deep Navajo Red stain or a semitransparent stain, or allow the wood to weather to a sun-bleached grey before installation. These decisions are best made prior to selecting and installing your solid lumber paneling (Table 1-3).

STAINING AND FINISHING WOOD

Chapter 6 will offer a complete discussion of the types and methods of finishing solid lumber paneling. Some elements of staining and finishing wood should, however, be considered prior to selecting and installing your solid lumber paneling. Let's discuss them.

Wood is such a widely available material and so easily adapted to man's use that he learned how to preserve it, and incidentally beautify it, long before the advent of written history. Paradoxically, as wood comes from nature, so it must be protected from nature. Climate and insects both take their toll of wood, indoors and out.

In general wood must be preserved from three things: fungus, insects, and moisture. Fungus and insects attack wood, drill it full of holes, reduce it to sponge, or turn it into a powder. The treatment for them is an application of copper napthenate or other preservative. Water or moisture also harms wood. In order to prevent this damage, wood is either painted, or a waterproof surface, called a *finish*, is applied to it. If in applying a protective surface you can also improve the appearance of the wood, why not do so?

Much of the beauty of wood lies in its color and its markings, which are also called *grain* and *figure*, respectively. Many woods, especially the fancy tropical woods, have a color that is not improved by staining. Yet even these woods don't reveal their true beauty until they have been smoothed, coated,

Table 1-3. Colors and Color Characteristics.

Reds:
1. English vermilion ....Bright, clear, slightly orange-red
2. Turkey red ....Bright, clear, slightly less intense
3. Permanent red ....Bright, clear, fire-engine red
4. Venetian red ....Brick red, dull
5. Indian red ....Bluer than Venetian, dull
6. Rose pink ....Purplish red, bright

Yellows:
1. Chrome yellow medium ....Clear, egg-yolk hue
2. Chrome yellow orange ....Clear, orange-yellow
3. Chrome yellow light ....Clear, greenish, light
4. Ochre (French or domestic) ....Dull, tan, wheat
5. Golden ochre ....Brighter and lighter than Ochre

Blues:
1. Ultramarine blue ....Clear, dark, purplish
2. Cobalt blue ....Clear, light, sky-blue
3. Prussian blue ....Clear, dark, greenish

Greens:
1. Chrome green medium ....Clear, dark, bluish
2. Chrome green dark ....Clear, darker, bluish
3. Chrome green light ....Clear, yellowish
4. Mitis green medium ....Clear, light, bright, yellowish

Browns:
1. Burnt umber ....Medium, clear, chocolate-brown
2. Raw sienna ....Light, tan, brown
3. Vandyke brown ....Dark reddish brown
4. Burnt sienna ....Very reddish brown
5. Raw umber ....Very grayish brown

Purple:
1. Madder lake ....Bright, dark, reddish

Black:
1. Lamp black ....Clear, sometimes greenish
2. Drop or ivory black..Clear, occasionally brownish

and polished. If this is true of these naturally beautiful woods, it is even more so of the paler and less vigorously figured woods of our temperate zone.

The art in staining and finishing lies in good judgment. The color must be strong enough to enhance the beauty of the wood without being too strong. Very dark staining obscures the delicate markings of the wood, toning down instead of emphasizing the natural textures. Stained wood is dull and doesn't brighten until the finish has been ap-plied. The finish clears and gives contrast to the grain and the stained wood. Different finishes can be gotten from a variety of materials. These materials include oils, waxes, shellacs, varnishes, and lacquers. In the order listed, they will provide a gloss from almost dull to very bright.

By carefully selecting the stain and the finish, and just as carefully applying them, you should obtain satisfactory results. The instructions offered in Chapter 6 will cover all the usual woods made into solid lumber paneling.

In addition to so-called brown staining, there is a blonde finish that is popular with interior decorators and furniture manufacturers. Then, too, there is the processing of less expensive woods to resemble the more valuable types.

Woods are commonly classified as hard or soft. This classification indicates the ability of the wood to absorb. In general, softwoods absorb the stain more readily, and consequently require a lighter mixture. American hardwoods include birch, cherry, hickory, maple, oak, and walnut. American-softwoods include cypress, fir, yellow pine, white pine, and redwood. Imported woods are Avodite, Circassian walnut, Philippine mahogany, true mahogany, rosewood, and satinwood. A great discussion of the various types of wood is offered in Chapter 2 Selecting Solid Lumber Paneling.

UNDERSTANDING GRAINS

Much of the character of wood as an element of interior decoration comes from the grain of the wood. In fact, the more expensive solid lumber paneling will be those with the most beautiful grains: rich and fluid. Grain is first formed by nature in the living wood. It is the result of the cells, fibers, and pores of the wood's structure; of the rings, streaks, veins, and density of the wood's growth depending on species, age, season, food, and climate; and of the twists and turns the wood develops, whether they arise from outside forces or from the meeting of trunk and branch and the formation of crotches.

While no two pieces of wood will be identical in their graining, each type has its own recognizable character. Douglas fir, for instance, has a very strong figure, which results from alternating bands of soft and hard wood. So strong is this figure, in fact, that this wood is not thought of as a "cabinet" or decorative wood. On the other hand, you are probably familiar with the warm color and close grain of cherry. An example of an exotic wood with an intricate pattern is Circassian walnut, which is vigorously gnarled and twisted by the weather of the Caucasus.

Grain in wood varies principally in the type of pore. The hardwoods—oak, ash, hickory, and

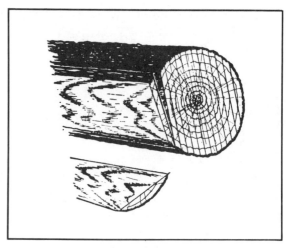

Fig. 1-28. Slash cutting proceeds straight across the log. This results in one kind of figure in the wood.

chestnut—resemble one another in the large opening pore characteristic of their grain. Softwoods—white pine, white wood, cedar, redwood, cypress, and some maples—resemble one another in the close, almost invisible, pores of the grain marking. Medium-sized pores are found in mahogany, walnut, and cherry. Each of these groups is grained differently.

Often we speak of the *formation* of grain, referring to the way in which wood is cut. Wood is cut in one of three ways. The most common is called *slash cut* (Fig. 1-28). Plank after plank is cut, starting on one side of the log and continuing to the

Fig. 1-29. Quarter cutting is like slash cutting except that the log is quartered first.

other side. In *quarter cutting* the log is cut into four equal wedge-shaped pieces. Then these wedges are cut into planks as shown in Fig. 1-29. *Veneers* can be cut the same as planks; they are simply cut much thinner. The third method of cutting is used exclusively for veneer. It is called *rotary cutting* (Fig. 1-30). The log is trued up and set on a massive machine resembling a lathe. A large knife is set against the revolving log and a thin sheet of wood is cut off. The method resembles pulling wrapping paper off a roll.

A different figure on the wood is made by each method of cutting. The first two are more commonly used for solid lumber paneling production, while the third—rotary cutting—is used for making plywood paneling.

Fig. 1-30. Rotary cutting is used to produce veneers.

PLANNING PANELING

As you can see, there are many considerations in planning your solid lumber paneling project: room size and style, colors, harmony, stains, finishes, wood grain, and type of paneling. In the coming chapters you will learn to select, prepare, install, and finish solid lumber paneling in your home or office for both beauty and function.

Chapter 2

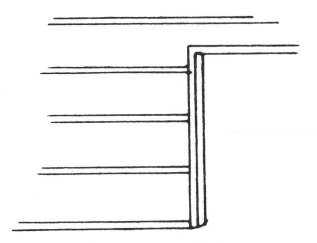

Selecting Solid Lumber Paneling

T HE LOOK OF WOOD CAN BE SIMULATED IN many ways, but real wood has a character of its own that offers both beauty and function to your home. More home decorators are choosing solid lumber paneling over simulated wood finishes for covering walls and other surfaces (Fig. 2-1 through 2-5). In this chapter, we'll look at the elements to be considered when selecting solid lumber paneling: the properties of wood and lumber, the types of solid lumber paneling, joints, and patterns.

LUMBER

Lumberyards and building material departments stock wood in two forms: rough and surfaced. *Rough lumber* is in the form in which it comes from the sawmill. The edges may not be square, and the surface is rough and shows saw marks. *Surfaced lumber* is planed on two or more sides and has square edges. Lumber is measured in the rough state, therefore surfaced lumber will measure slightly less, because of planing. In most cases the surfaced lumber will be about 1/4 inch under the rough. Thus, a piece of 2-×-4-inch stock will be about 1 5/8 × 3 5/8 inches when surfaced on all sides.

You'll find surfaced lumber best for practically every project (Fig. 2-6). It's much easier to work with, particularly when all sides are square. The planing required to make surfaced lumber out of rough takes only a short time with the right equipment. In fact, it's becoming more difficult to find rough lumber today, although it is used for some solid lumber paneling installations.

Secondhand lumber can be purchased at many yards and has many uses, especially in interior decorating. Remember, however, that used lumber may not be as strong as new wood and consequently should be limited to those jobs in which there is no great strain, such as wall paneling. Be very careful when working with used lumber not to strike a nail with the saw or plane. Examine the lumber carefully before working with it, watching for nails that have lost their heads and are difficult to find.

Fig. 2-1. Decorating a living room wall with solid lumber paneling (courtesy Georgia-Pacific).

Fig. 2-2. Sectional diagonal solid lumber paneling (courtesy Georgia-Pacific).

Fig. 2-3. Basement recreation room before solid lumber paneling (courtesy Georgia-Pacific).

Fig. 2-4. Solid lumber paneling installed in basement rec room (courtesy Georgia-Pacific).

Fig. 2-5. One rec room wall covered with solid lumber paneling (courtesy Georgia-Pacific).

MEASURING LUMBER

Lumber comes in many different standard lengths, running to about 18 feet. The widths vary from 2 inches to 12 and thickness from 1 to 8 inches (Table 2-1).

The boardfoot is the standard measurement for lumber. This unit of measurement is equal to a board 1 inch thick × 12 inches wide × 12 inches long. To find the number of board feet in a piece of lumber, multiply the length in feet by the width and thickness in inches and divide by 12. As an example, a 2- × -4- × -8-foot board is 5 1/3 board feet (2 × 4 = 8 × 8 = 64 ÷ 12 = 5.33).

You'll find that many lumberyards will quote the price of lumber in board feet. It's best, therefore, to know how many board feet you need for your paneling or other project.

When lumber is first cut, it contains a considerable amount of moisture. Under normal circumstances lumber is stacked in piles after it has been sawed and left to dry. During this period most of the moisture will evaporate, but a small portion will remain, and this may require several months or more to disappear. Still-moist lumber is called *green* lumber. After most of the moisture has evaporated, the lumber is called *seasoned*. Green lumber is not suited to most building needs, including paneling, because it shrinks as it dries. To build a house or cover a wall with green lumber is to invite problems.

To hasten the drying of the wood, green lumber is often put in an oven called a *kiln*. This is an artificial method of seasoning the wood and, consequently, makes the wood more expensive; however kiln-dried wood will have a more uniform moisture content than air-dried wood in many cases.

WOODS

Knowing the important characteristics of the popular kinds of wood is helpful in selecting solid

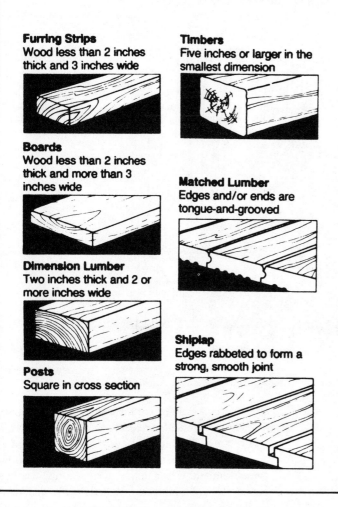

Furring Strips
Wood less than 2 inches thick and 3 inches wide

Timbers
Five inches or larger in the smallest dimension

Boards
Wood less than 2 inches thick and more than 3 inches wide

Matched Lumber
Edges and/or ends are tongue-and-grooved

Dimension Lumber
Two inches thick and 2 or more inches wide

Posts
Square in cross section

Shiplap
Edges rabbeted to form a strong, smooth joint

Fig. 2-6. Numerous types of lumber available through neighborhood building materials firms (courtesy Georgia-Pacific).

lumber paneling. Unfortunately, what is called one type of wood in one region may be called something else in another. Let's consider the properties of common wood species used in home construction and paneling (Table 2-2).

Spruce is a close-grain wood that is used chiefly for framing houses. It is light in color and easy to work. While strong enough for framing purposes, spruce doesn't stand up very well when exposed to the weather and shouldn't be used for exterior work. It is often used in manufacturing solid lumber paneling for interior installation.

There are two types of *fir*, the Eastern and the Douglas. Fir is used extensively for framing and for floors and trim. It is light in weight, easy to work, and is used a great deal in the manufacture of plywood. A close-grain wood, fir takes both stain and paint well.

White pine at one time was used for almost all home construction. Now it is used primarily for trim and some paneling. Light in weight, white pine is easy to work and has a close grain.

Yellow pine is also called hard pine or by a number of other names. It is strong enough to be used for framing and comes in sufficiently wide boards to be used for paneling. It's easy to work, but rather difficult to paint.

Hemlock is not very suitable for the majority

Table 2-1. Nominal and Dressed Sizes of Lumber.

ITEM	THICKNESSES		FACE WIDTHS	
	NOMINAL	DRESSED Inches	NOMINAL	DRESSED Inches
Boards	1 1-1/4 1-1/2	3/4 1 1-1/4	2 3 4 5 6 7 8 9 10 11 12 14 16	1-1/2 2-1/2 3-1/2 4-1/2 5-1/2 6-1/2 7-1/4 8-1/4 9-1/4 10-1/4 11-1/4 13-1/4 15-1/4
Dimension	2 2-1/2 3 3-1/2	1-1/2 2 2-1/2 3	2 3 4 5 6 8 10 12 14 16	1-1/2 2-1/2 3-1/2 4-1/2 5-1/2 7-1/4 9-1/4 11-1/4 13-1/4 15-1/4
Dimension	4 4-1/2	3-1/2 4	2 3 4 5 6 8 10 12 14 16	1-1/2 2-1/2 3-1/2 4-1/2 5-1/2 7-1/4 9-1/4 11-1/4
Timbers	5 & Thicker		5 & Wider	

Table 2-2. Common Woods and Their Uses. (Continued through page 32.)

Type	Sources	Uses	Characteristics
Ash	East of Rockies . .	Oars, boat thwarts, benches, gratings, hammer handles, cabinets, ball bats, wagon construction farm implements.	Strong, heavy, hard, tough, elastic, close straight grain, shrinks very little, takes excellent finish, lasts well.
Balsa	Ecuador.	Rafts, food boxes, linings of refrigerators, life preservers, loud speakers, sound-proofing, air-conditioning devices, model airplane construction.	Lightest of all woods, very soft, strong for its weight, good heat insulating qualities, odorless.
Basswood .	Eastern half of U.S. with exception of coastal regions.	Low-grade furniture, cheaply constructed buildings, interior finish, shelving, drawers, boxes, drainboards, woodenware, novelties, excelsior, general millwork.	Soft, very light, weak, brittle, not durable, shrinks considerably, inferior to poplar, but very uniform, works easily, takes screws and nails well and does not twist or warp.
Beech. . . .	East of Mississippi, Southeastern Canada.	Cabinetwork, imitation mahogany furniture, wood dowels, capping, boat trim, interior finish, tool handles, turnery, shoe lasts, carving, flooring.	Similar to birch but not so durable when exposed to weather, shrinks and checks considerably, close grain, light or dark red color.
Birch	East of Mississippi River and North of Gulf Coast States, Southeast Canada, Newfoundland.	Cabinetwork, imitation mahogany furniture, wood dowels, capping, boat trim, interior finish, tool handles, turnery, carving.	Hard, durable, fine grain, even texture, heavy, stiff, strong, tough, takes high polish, works easily, forms excellent base for white enamel finish, but not durable when exposed. Heartwood is light to dark reddish brown in color.
Butternut	Southern Canada, Minnesota, Eastern U. S. as far south as Alabama and Florida.	Toys, altars, woodenware, millwork, interior trim, furniture, boats, scientific instruments.	Very much like walnut in color but softer, not so soft as white pine and basswood, easy to work, coarse grained, fairly strong.

Type	Sources	Uses	Characteristics
Cypress....	Maryland to Texas, along Mississippi valley to Illinois.	Small boat planking, siding, shingles, sash, doors, tanks, silos, railway ties.	Many characteristics similar to white cedar. Water resistant qualities make it excellent for use as boat planking.
Douglas Fir..	Pacific Coast, British Columbia.	Deck planking on large ships, shores, strongbacks, plugs, filling pieces and bulkheads of small boats, building construction, dimension timber, plywood.	Excellent structural lumber, strong, easy to work, clear straight grained, soft, but brittle. Heartwood is durable in contact with ground, best structural timber of northwest.
Elm.......	States east of Colorado.	Agricultural implements, wheel-stock, boats, furniture, crossties, posts, poles.	Slippery, heavy, hard, tough, durable, difficult to split, not resistant to decay.
Hickory.....	Arkansas, Tennessee, Ohio, Kentucky.	Tools, handles, wagon stock, hoops, baskets, vehicles, wagon spokes.	Very heavy, hard, stronger and tougher than other native woods, but checks, shrinks, difficult to work, subject to decay and insect attack.
Lignum Vitae	Central America.	Block sheaves and pulleys, waterexposed shaft bearings of small boats and ships, tool handles, small turned articles, and mallet heads.	Dark greenish brown, unusually hard, close grained, very heavy, resinous, difficult to split and work, has soapy feeling.
Live Oak ...	Southern Atlantic and Gulf Coasts of U.S., Oregon, California.	Implements, wagons, ship building.	Very heavy, hard, tough, strong, durable, difficult to work, light brown or yellow sap wood nearly white.
Mahogany ...	Honduras, Mexico, Central America, Florida, West Indies, Central Africa, other tropical sections.	Furniture, boats, decks, fixtures, interior trim in expensive homes, musical instruments.	Brown to red color, one of most useful of cabinet woods, hard, durable, does not split badly, open grained, takes beautiful finish when grain is filled but checks, swells, shrinks, warps slightly.

Type	Sources	Uses	Characteristics
Maple.....	All states east of Colorado, Southern Canada.	Excellent furniture, high-grade floors, tool handles, ship construction crossties, counter tops, bowling pins.	Fine grained, grain often curly or "Bird's Eyes," heavy, tough, hard, strong, rather easy to work, but not durable. Heartwood is light brown, sap wood is nearly white.
Norway Pine.....	States bordering Great Lakes.	Dimension timber, masts, spars, piling, interior trim.	Light, fairly hard, strong, not durable in contact with ground.
Philippine Mahogany..	Philippine Islands	Pleasure boats, medium-grade furniture, interior trim.	Not a true mahogany, shrinks, expands, splits, warps, but available in long, wide, clear boards.
Poplar	Virginias, Tennessee, Kentucky, Mississippi Valley.	Low-grade furniture cheaply constructed buildings, interior finish, shelving, drawers, boxes.	Soft, cheap, obtainable in wide boards, warps, shrinks, rots easily, light, brittle, weak, but works easily and holds nails well, fine-textured.
Red Cedar..	East of Colorado and north of Florida.	Mothproof chests, lining for linen closets, sills, and other uses similar to white cedar.	Very light, soft, weak, brittle, low shrinkage, great durability, fragrant scent, generally knotty, beautiful when finished in natural color, easily worked.
Red Oak...	Virginias, Tennessee, Arkansas, Kentucky, Ohio, Missouri, Maryland.	Interior finish, furniture, cabinets, millwork, crossties when preserved.	Tends to warp, coarse grain, does not last well when exposed to weather, porous, easily impregnated with preservative, heavy, tough, strong.
Redwood ..	California.	General construction, tanks, paneling.	Inferior to yellow pine and fir in strength, shrinks and splits little, extremely soft, light, straight grained, very durable, exceptionally decay resistant.

Type	Sources	Uses	Characteristics
Spruce	New York, New England, West Virginia, Central Canada, Great Lakes States, Idaho, Washington, Oregon.	Railway ties, resonance wood, piles, airplanes, oars, masts, spars, baskets.	Light, soft, low strength, fair durability, close grain, yellowish, sap wood indistinct.
Sugar Pine	California, Oregon.	Same as white pine.	Very light, soft, resembles white pine.
Teak	India, Burma, Siam, Java.	Deck planking, shaft logs for small boats.	Light brown color, strong, easily worked, durable, resistant to damage by moisture.
Walnut	Eastern half of U.S. except Southern Atlantic and Gulf Coasts, some in New Mexico, Arizona, California.	Expensive furniture, cabinets, interior woodwork, gun stocks, tool handles, airplane propellers, fine boats, musical instruments.	Fine cabinet wood, coarse grained but takes beautiful finish when pores closed with woodfiller, medium weight, hard, strong, easily worked, dark chocolate color, does not warp or check, brittle.
White Cedar. . . .	Eastern Coast of U.S., and around Great Lakes.	Boat planking, railroad ties, shingles, siding, posts, poles.	Soft, light weight, close grained, exceptionally durable when exposed to water, not strong enough for building construction, brittle, low shrinkage, fragment, generally knotty.
White Oak . .	Virginias, Tennessee, Arkansas, Kentucky, Ohio, Missouri, Maryland, Indiana.	Boat and ship stems, sternposts, knees, sheer strakes, fenders, capping, transoms, shaft logs, framing for buildings, strong furniture, tool handles, crossties, agricultural implements, fence posts.	Heavy, hard, strong, medium coarse grain, tough, dense, most durable of hardwoods, elastic, rather easy to work, but shrinks and likely to check. Light brownish grey in color with reddish tinge, medullary rays are large and outstanding and present beautiful figures when quarter sawed, receives high polish.

Type	Sources	Uses	Characteristics
White Pine.	Minnesota, Wisconsin, Maine, Michigan, Idaho, Montana, Washington, Oregon, California	Patterns, any interior job or exterior job that doesn't require maximum strength, window sash, interior trim, millwork, cabinets, cornices.	Easy to work, fine grain, free of knots, takes excellent finish, durable when exposed to water, expands when wet, shrinks when dry, soft, white, nails without splitting, not very strong, straight grained.
Yellow Pine.	Virginia to Texas.	Most important lumber for heavy construction and exterior work, keelsons, risings, filling pieces, clamps, floors, bulkheads of small boats, shores, wedges, plugs, strongbacks, staging, joists, posts, piling, ties, paving blocks.	Hard, strong, heartwood is durable in the ground, grain varies, heavy, tough, reddish brown in color, resinous, medullary rays well marked.

of home construction applications because it shrinks and splits easily. Hemlock is brown in color, holds nails very well, and can be used for framing, sheathing, and other rough construction work, but is usually not used for solid lumber paneling. An open-grain wood, hemlock takes paint poorly.

Known chiefly for the pleasant effect obtained when used for paneling, *cypress* can be used for interior trim. It is very resistant to weather and, consequently, is suitable for outside fittings such as siding. The wood has a close grain and will take stain well. It absorbs paint slowly; so ample time should be allowed between coats.

A heavy and difficult wood to work, *oak* is used for better solid lumber paneling and for trim. It's also popular for building cabinets and floors. Oak is open-grained and needs a filler. It can be painted, but absorbs the paint slowly.

Birch is frequently used for interior trim and furniture because it is hard and strong, but it doesn't stand outside exposure very well. Birch is easy to work and is close-grained.

Walnut is a very expensive wood, and it is not used for solid lumber paneling for most homes for this reason. A walnut veneer, however, can be a beautiful addition to a wall or cabinet. American walnut is easy to work and has an open grain.

Red cedar is used for exterior work because of its ability to stand up to weather, and for interior work because of its beauty. It has a close grain and withstands exposure well, but it cannot be painted until it is well seasoned.

Ash is a tough, elastic hardwood sometimes used for trim. It is open-grained, requires a filler, and takes paint well.

Redwood can be used for all types of building work. It is a light wood and easy to work. Close-grained, it takes paint readily and can be stained and polished. It is very popular as solid lumber paneling because of its beauty and workability.

Elm is tough, hard, and damp resistant. It is used mostly for heavy timbers and framework. It is difficult to work and has an open grain.

Maple is a very hard and strong wood used for flooring and for trim. It is difficult to nail. Maple is close-grained and takes paint very well.

Mahogany is used chiefly for furniture and some paneling. This wood is very strong and easy to work. It has an open grain.

Light in weight and fairly strong, *poplar* is used for interior trim, including solid lumber paneling. It can be stained and polished to resemble more expensive kinds of wood. Poplar doesn't stand exposure well, but it is easy to paint.

LUMBER SIZES

Softwood lumber is classified by its nominal size as boards, dimensions, and timbers. *Boards* are less than 2 inches in thickness and 2 or more inches in width. Those less than 6 inches in width may be classified as *strips*. *Dimensions* are from 2 inches to, but not including, 5 inches in thickness and 2 inches or more in width. Joists, planks, rafter, and studs are examples of dimension lumber. Solid lumber paneling is classified as boards. *Timbers* are 5 or more inches in both thickness and width. Beams, girders, posts, caps, etc., are classified as timbers.

GRADING LUMBER

Softwood lumber is graded for quality in accordance with American Lumber Standards. The major quality grades are select lumber and common lumber. Select lumber is of better quality than common lumber. Each of these grades has subdivisions in descending order of quality as follows:

☐ Grade A lumber is select lumber which is practically free of defects and blemishes.

☐ Grade B lumber is select lumber which contains a few minor blemishes.

☐ Grade C lumber is finish lumber which contains more numerous and more significant blemishes than grade B. It must be capable of being easily and thoroughly concealed with paint.

☐ Grade D lumber is finish lumber which contains more numerous and more significant blemishes than grade C, but which is still capable of presenting a satisfactory appearance when painted.

☐ No. 1 common lumber is sound, tightknotted stock, containing only a few minor defects. Lumber in this classification must be suitable for use as watertight lumber.

☐ No. 2 common lumber contains a limited number of significant defects, but no knotholes or other serious defects. Lumber classified as No. 2 common must be suitable for using as graintight lumber.

☐ No. 3 common lumber contains a few defects which are larger and coarser than those in No. 2 common; occasional knotholes, for example.

☐ No. 4 common lumber is low-quality material containing serious defects like knotholes, checks, shakes, and decay.

☐ No. 5 common is capable only of holding together under ordinary handling.

Though you will probably want the highest grade you can afford for your solid lumber paneling, you may want to look at even the lowest grades for their character if they don't need to have extensive strength.

WORKING WITH PLYWOOD

While this is a book on solid lumber and solid lumber paneling, plywood cannot be ignored. It is a primary building material that may be used as subflooring, rough wall surfacing, and other functions. Even some strip paneling is made of plywood.

Plywood is a panel product made from thin sheets (plies) of wood, called *veneers,* which are laminated together. The grain of each ply normally runs at right angles to adjacent plies (Fig. 2-7). An odd number of veneers—three, five, or seven—is generally used so the grain direction on the face and back of the panel run in the same direction. Cross-lamination distributes the grain strength in both directions, creating a panel that is split-proof and pound for pound one of the strongest building materials available. Plywood can be worked quickly and easily with common carpentry tools. It holds nails well and normally doesn't split when nails are driven close to the edges. Finishing plywood presents no unusual problems; it may be sanded or

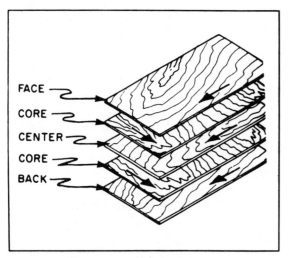

Fig. 2-7. Plies of wood in plywood.

crements. Other sizes of plywood are available on special order. The thickness of plywood usually runs from 1/4 to 3/4 inch, but other sizes and thickness may be obtained.

GRADING PLYWOOD

All plywood panels are graded as to quality, based on product standards. The grade of each type of plywood is determined by the kind of veneer (N, A, B, C, or D) used for the face and back of the panel (Fig. 2-8) and also by the type of glue used in construction. For example, a sheet of plywood having the designation A-C (Fig. 2-9) will have the A-grade veneer on the face and the C-grade veneer on the back. Grading is also based on the number of defects, such as knotholes, pitch pockets, splits, discolorations, and patches in the face of each panel. Each panel or sheet of plywood has a stamp on the back that gives all the information you will need.

Plywood is classified into one of two types: interior or exterior, depending on their capability to withstand weather exposure. The type also denotes veneer grade and adhesive durability. There are a number of grades within each type of plywood, based on the quality of the veneer of the panel.

Interior-type plywood will withstand an occasional wetting during construction, but should not be permanently exposed to the elements. Within

texture-coated with a permanent finish, or it may be left to weather naturally.

There is probably no building material as versatile as plywood. It is used for concrete forms, wall and roof sheathing, flooring, box beams, soffits, stressed-skin panels, paneling, shelving, doors, furniture, cabinets, crates, signs, and many other items.

Plywood panels commonly used in building construction come in standard sizes of 4 × 8 feet or 48 × 96 inches. Plywood is available, however, in panel widths of 36, 48, and 60 inches. Panel lengths range from 60 to 144 inches in 12-inch in-

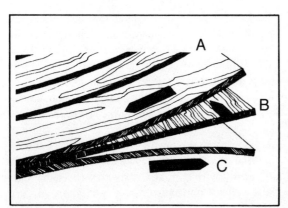

Fig. 2-8. Face-and-back ply grading of plywood (courtesy Georgia-Pacific).

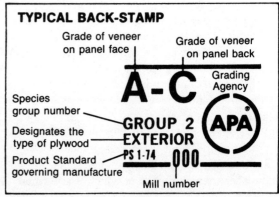

Fig. 2-9. Typical plywood grade back-stamp (courtesy Georgia-Pacific).

34

the interior-type classification, there are three levels of adhesive durability:

1 - interior with interior glue, which may be used where the plywood will not be subject to continuing moisture conditions or extreme humidity

2 - interior with intermediate glue which is bonded with adhesives possessing high-level resistance to bacteria, mold, and moisture

3 - interior with exterior waterproof glue

Because of the roughness of the inner plies of interior plywood, these panels are not equal in durability to exterior plywood.

Exterior-type plywood is produced with C-grade veneers or better throughout and bonded thoroughly with a waterproof adhesive. It retains the glue bond when wet and dried repeatedly or otherwise subjected to the weather. It is intended for permanent exterior exposure.

UNDERSTANDING CABINETRY

Solid lumber paneling can be installed on cabinets and furniture, as well as walls. Even in wall installation, however, you will need to understand cabinet-making and millwork as you install molding and trim. Let's take a quick look at these subjects.

Figure 2-10 illustrates simple molding and trim shapes used to decorate walls and other surfaces covered with solid lumber paneling. The term *contour cutting* refers to the cutting of ornamental face curves on wood stock which is used for molding and other trim. Some moldings and trims are purchased at building material outlets. Others can be done at home on a shaper equipped with a cutter or blades, or with a combination of cutters and/or blades, arranged to produce the desired contour. Figure 2-11 shows the simplest and most common moldings and trims used in woodworking.

As a general term, *millwork* usually embraces most wood products and components that require manufacturing. It not only includes the interior trim and doors, but also kitchen cabinets and similar units that are often covered with solid lumber paneling after installation. Most of these units are pro-

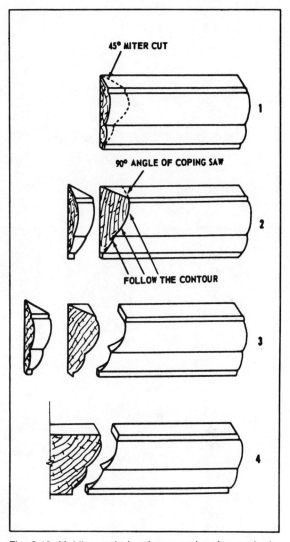

Fig. 2-10. Molding and trim shapes and coping methods.

duced in a millwork manufacturing plant and are ready to install. They usually require only fastening to the wall or floor. Figures 2-12 and 2-13 show typical types of millwork.

The construction details for various kinds of furniture and cabinets are similar. Dressers, chests of drawers, kneehole desks, and built-in cabinets all have drawers for storage purposes and are constructed in a similar way.

A number of pieces of stock glued edge to edge should provide sufficient width for the cabinet side.

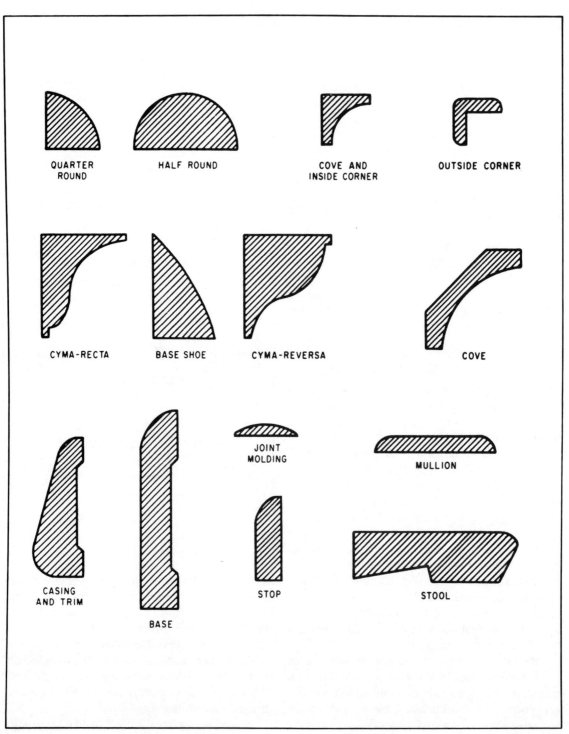

Fig. 2-11. Common moldings used in woodworking.

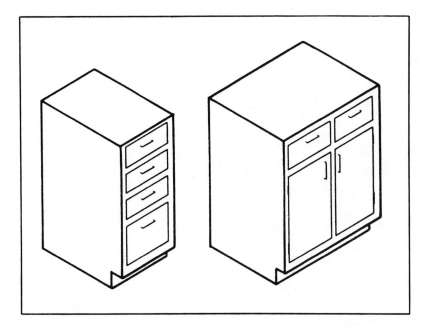

Fig. 2-12. Floor cabinets.

The sides usually range from 16 to 20 inches and are finished 3/4 inch thick. Some constructions require the use of a square post in each corner of the case. If the square posts are used, they are usually connected with the rails.

Division rails and bearing rails are used to make up the drawer division frames. These frames are usually made of stock 3/4 inch thick and 2 inches wide and fastened together at the corners with blind mortise-and-tenon joints. The frames are glued and checked to ensure they are square and the same size. Usually a gain is cut on the front edge of the frames, as shown in Fig. 2-14, and fitted and glued into the dadoes on the inside faces of the sides. The frames should be made 1/4 inch narrower than the sides when 1/4-inch plywood is used to allow the

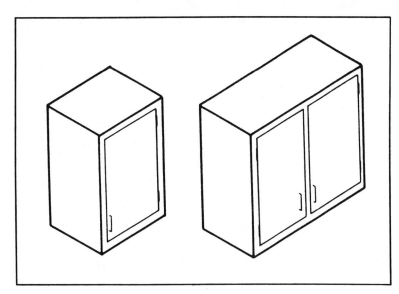

Fig. 2-13. Wall cabinets.

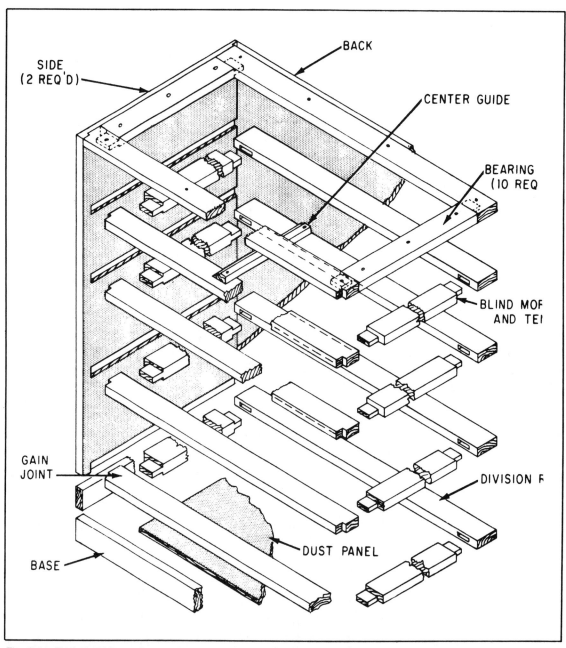

Fig. 2-14. Typical cabinet components.

plywood back to be glued and fastened into place. To keep dust and insects out of the case, a plywood panel can be installed in the lower frame.

Drawers require special attention so they fit ac-

curately. Drawers have been known to expand or shrink in the front, sides, and back. Therefore, some allowance must be made accordingly. It will be helpful to you if you remember that wood shrinks or expands mostly across the grain and very little lengthwise or with the grain. The height of the drawer opening should be 1/8 to 3/16 inch wider than the drawer sides and back. The edges and ends should be slightly beveled toward the back face and the drawer front fitted to the opening. The lip-type drawer may be fitted more loosely because the lip extension will cover the opening at the end and top. There are several types of joints that can be used to join the drawer sides to the front: plain rabbet joint, half blind dovetail, and the dado tongue and rabbet.

The bottom of the drawer is usually grooved into all four sides of the drawer, and a plain grooved or dovetail center guide is fastened to the bottom. After the drawer has been fitted into place and the front lined up with the front face of the case, the center guide is fastened permanently to the drawer rails with screws. You may find it more convenient if you leave the back off the case until the drawer guides have been fastened to the drawer rails.

Solid lumber paneling can then be laid over the surface of the cabinet, top, and sides (Fig. 2-15). You can also replace the plywood top and sides with solid lumber paneling as you build the cabinet (Figs. 2-16 and 2-17).

To minimize expansion and shrinkage of the wood, a sealer coat of finish should be applied to the inside and outside surfaces of the case and its drawers. It will be discussed in greater detail in Chapter 6 Finishing.

JOINTS

There are many types and designs of joints used in milling and installing solid lumber paneling. The most common joints for solid lumber paneling follow.

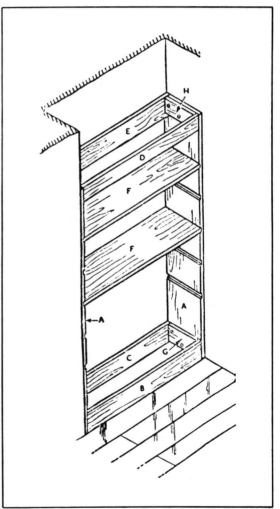

Fig. 2-16. Solid lumber paneling can be used to make your own built-in cabinet.

Fig. 2-15. Cabinets can be faced with solid lumber paneling for a rich effect (courtesy Georgia-Pacific).

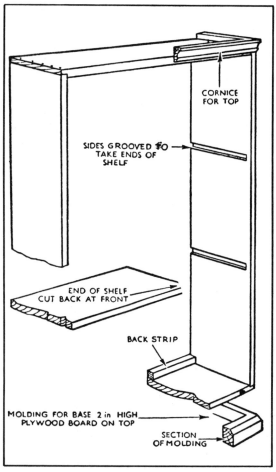

SIDES GROOVED TO TAKE ENDS OF SHELF

CORNICE FOR TOP

END OF SHELF CUT BACK AT FRONT

BACK STRIP

MOLDING FOR BASE 2 in HIGH PLYWOOD BOARD ON TOP

SECTION OF MOLDING

Fig. 2-17. Details of shelving made from solid lumber paneling.

Figure 2-18 illustrates the plain *butt joint,* which is simply butting two pieces of wood together tightly. It is the most economoical joint because it doesn't require millwork and can be installed with finished stock from any lumberyard.

Figure 2-19 shows the most common joint used in milling and installing solid lumber paneling—the *tongue-and-groove joint.* It is easy to mill and easy to install, and gives a tight fit that won't come apart as easily as the plain butt joint (Figs. 2-20 through 2-22).

There are many variations to the popular tongue-and-groove joint, including the *spline joint* (Fig. 2-23). For the spline joint, a groove is cut in two pieces of wood, and a tongue board or spline is inserted between them. While it is more difficult to install than prepared tongue-and-groove paneling, it can be milled more easily in your shop. Figure 2-24 shows the related *doweled joint.*

Another type of solid lumber paneling joint is the *board and batten* (Fig. 2-25). As you can see, it is installed by simply spacing the boards as you install them on the wall, then covering the spaces with narrow boards, or *battens.*

A variation of the board-and-batten joint is called the *board-and-board joint* a very descriptive name. The space between the first row of boards is slightly less than the width of a board. The second row of boards overlaps this space. Board-and-board joints are most popular with seasoned or salvaged wood, such as that gathered from the

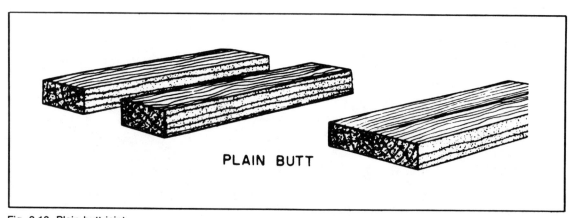

PLAIN BUTT

Fig. 2-18. Plain butt joint.

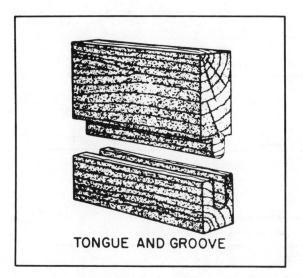

TONGUE AND GROOVE

Fig. 2-19. Tongue-and-groove joint.

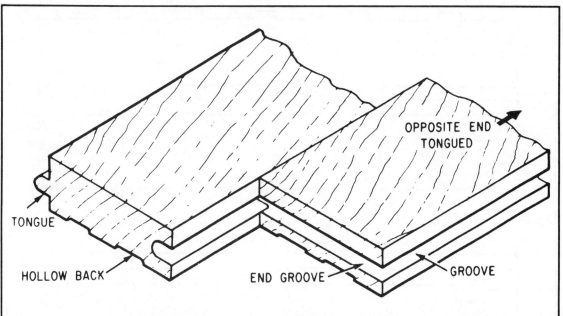

TONGUE

HOLLOW BACK

OPPOSITE END
TONGUED

END GROOVE

GROOVE

Fig. 2-20. Note how the tongue fits tightly into the groove to hold the two pieces of wood more firmly together than the plain butt joint.

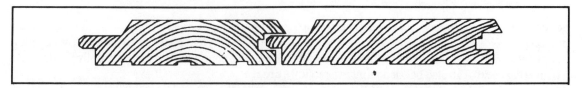

Fig. 2-21. Side view of typical tongue-and-groove solid lumber paneling (courtesy Potlatch).

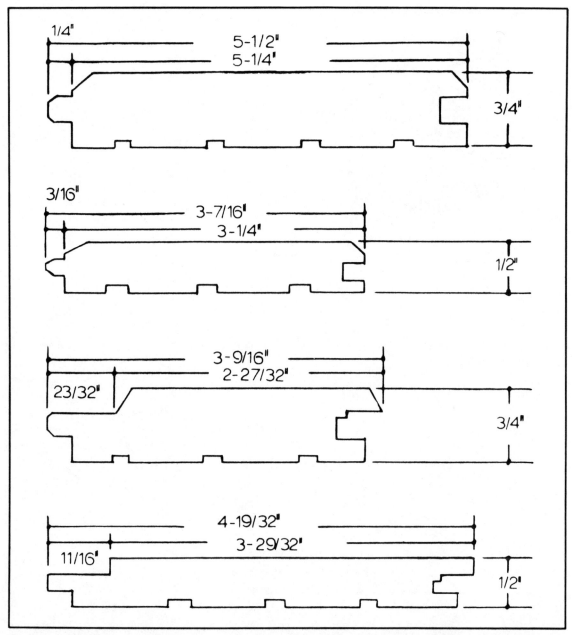

Fig. 2-22. Configurations and dimensions of typical tongue-and-groove solid lumber paneling (courtesy Potlatch).

demolition of a barn or other weathered outbuilding.

Bevel siding, often installed on the exterior of a home (Fig. 2-26), can also be used as interior solid lumber paneling. In fact, because of the wide use and production of exterior siding, there are many types and styles available to the do-it-yourselfer, including common bevel, anzac, dolly varden, and drop.

Finally, *channel* (Fig. 2-27) solid lumber panel-

42

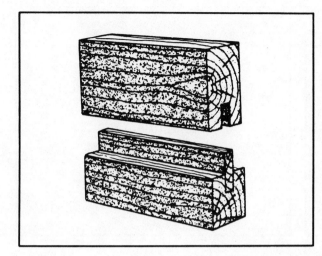

Fig. 2-23. Spline joint.

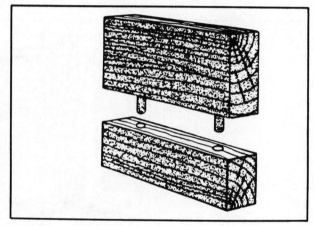

Fig. 2-24. Doweled joint.

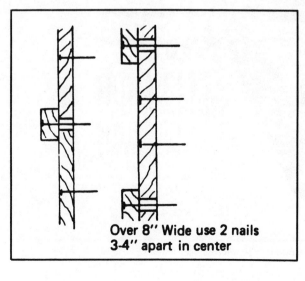

Over 8″ Wide use 2 nails 3-4″ apart in center

Fig. 2-25. Board-and-batten joint.

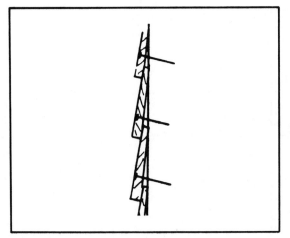

Fig. 2-26. Bevel paneling or siding.

Fig. 2-27. Channel paneling.

Fig. 2-28. Solid lumber paneling can enhance furniture (courtesy Georgia-Pacific).

44

Fig. 2-29. Paneling in a basement room (courtesy Georgia-Pacific)

ing is very popular with many do-it-yourselfers because it offers the milled look of tongue-and-groove paneling with easier installation. Channel paneling can be quickly set into place and nailed, making a wall or other surface decorative in a few hours.

BUYING SOLID LUMBER PANELING

Once you've decided on the type of wood, its use, and the method of installation, you'll be ready to purchase it. The best and least costly way of doing so is to "bid it out;" that is, you write down the specifications—including species, size, milling, grade, and amount—and take it to a few sources, such as building material outlets and lumberyards. By doing so you can sit down at home without the pressure of salesmanship and compare apples with apples. That is, you can decide which supplier has the greatest offer.

Price is only one aspect of the greatest offer. The terms, the location, add-ons, taxes, and other elements also enter into the cost. One place may

Fig. 2-30. Combination of wall and ceiling solid lumber paneling (courtesy Georgia-Pacific).

have a higher bid, but offer to deliver the materials. If you don't have a vehicle that will hold the paneling in one or two trips, this may be a better bargain than the seemingly lower price.

Once you have selected and purchased your solid lumber paneling, it's time to prepare it for installation (Fig. 2-28 through 2-30). This subject is discussed in Chapter 3.

Chapter 3

Preparing Paneling

N OW THAT YOU'VE PLANNED AND SELECTED your solid lumber paneling, you can begin preparing the job—getting the paneling ready for the installation surface and vice versa. Before you install solid lumber paneling, you will want to make sure that the wall or other surface is finished and that it is adequately insulated, covered, and caulked. You will also want to prepare the panels for installation. First, though, let's take a look at the tools you will need for the job.

SELECTING TOOLS

The tool that will last (almost) forever is a fine bargain even though its initial cost will be more than what's displayed in an anything-for-a-dollar bin. Aside from the economics, the good tool helps make you a better craftsperson (Fig. 3-1). For instance, a saw of inferior metal with burred teeth that are incorrectly set will cut wood. Accuracy will be difficult to achieve, however, and the sawed edge will require much additional attention to make it acceptable.

Tools beyond the basic assortment recommended in this chapter should be acquired as the need for them arises, but only if they will be used often. There are many tools that fall into the rarely used category—nice to have, but an indulgence if they will merely gather dust. Here are some suggestions for must-have tools for the person planning, preparing, and installing solid lumber paneling and related finish work.

Basic Hand Tools

The *crosscut saw* (Fig. 3-2) has small teeth with knife like points. It is designed for cutting across the grain of lumber, but it is also good for sawing all types of plywood (Fig. 3-3). A model that is *taper ground*, which usually indicates a quality product, has 8 or 10 teeth per inch and is 26 inches long.

The *keyhole* and *compass saws* (Fig. 3-4) are different tools, but overlap in function. They have narrow blades that taper to a small point and are useful for sawing curved lines and for making internal cutouts; for example, an opening through a panel

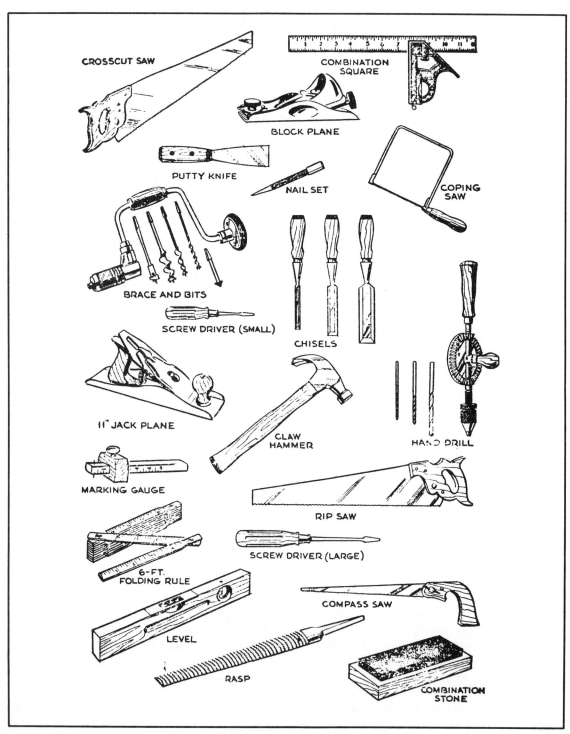

CROSSCUT SAW

COMBINATION SQUARE

BLOCK PLANE

PUTTY KNIFE

NAIL SET

COPING SAW

BRACE AND BITS

SCREW DRIVER (SMALL)

CHISELS

HAND DRILL

11" JACK PLANE

CLAW HAMMER

MARKING GAUGE

RIP SAW

6-FT. FOLDING RULE

SCREW DRIVER (LARGE)

COMPASS SAW

LEVEL

RASP

COMBINATION STONE

Fig. 3-1. Assortment of good tools for the craftsperson.

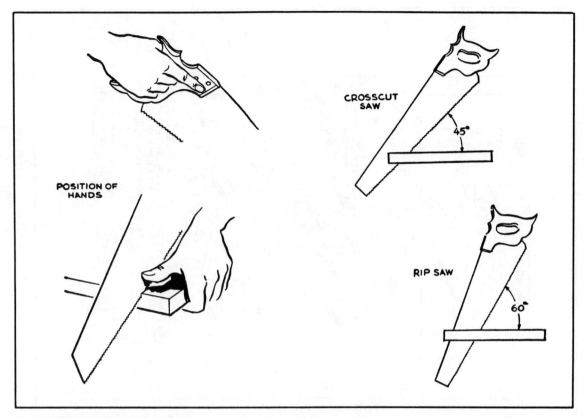

POSITION OF HANDS

CROSSCUT SAW

45°

RIP SAW

60°

Fig. 3-2. Saws and how to use them.

for an electrical outlet box. A single handle with three different blades is often available as a nest of saws, one of which can be used to saw metal.

The *backsaw* has teeth like a crosscut saw and a rectangular blade stiffened with a length of steel or brass along its top edge. It's for more precise work than can be accomplished with other saws. For example, it can be used for making the miter cut that is required when two pieces of molding are joined to make a 90-degree turn. It is often used with a *miter box* (Fig. 3-5), which is simply a guide for accurate cutting.

A 16-ounce *claw hammer* (Fig. 3-6) is for general use when driving or pulling out nails. The striking surface should be slightly convex (bell-faced) so you can drive a nail flush without damage to adjacent surfaces. The handle may be of hardwood, rubber-sheathed fiberglass, or steel. The choice is a personal one. Some old timers say wood

because it doesn't get cold to the touch in inclement weather. An argument for steel is that it will stay tightly fixed in the head regardless of moisture and changes in temperature.

Nail sets (Fig. 3-7) are a must because you often drive a finishing nail below the surface of the wood so it can be hidden with wood dough. A single nail set won't do, however. Nails come in different sizes; so nail sets are made to match. For the minimum, have sizes 1/32 and 3/32 inch.

It's good to have a *flex tape* (Fig. 3-8), but it must be at least an 8 footer since that's the longest dimension of a wall. They come in different widths—3/4 inch is a good choice, even though it is bulkier than others, because it will have the rigidity to span openings without buckling. Markings should include inches and feet and special indications for 16-inch on-center stud placement. A lock that will hold the tape at any extended posi-

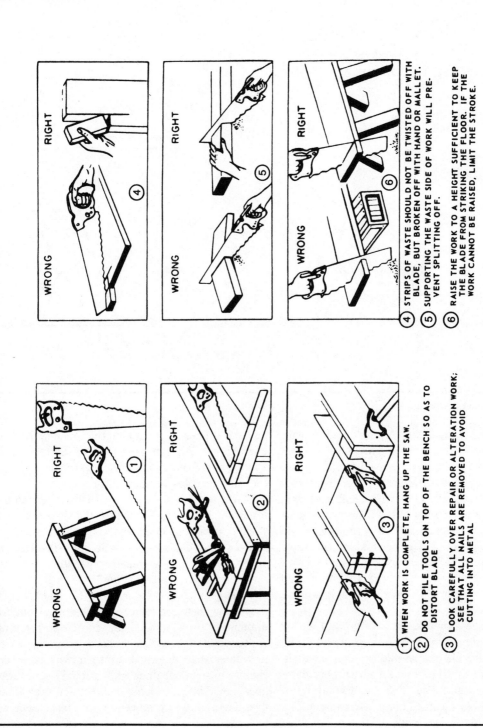

① WHEN WORK IS COMPLETE, HANG UP THE SAW.

② DO NOT PILE TOOLS ON TOP OF THE BENCH SO AS TO DISTORT BLADE

③ LOOK CAREFULLY OVER REPAIR OR ALTERATION WORK; SEE THAT ALL NAILS ARE REMOVED TO AVOID CUTTING INTO METAL

④ STRIPS OF WASTE SHOULD NOT BE TWISTED OFF WITH BLADE, BUT BROKEN OFF WITH HAND OR MALLET.

⑤ SUPPORTING THE WASTE SIDE OF WORK WILL PRE- VENT SPLITTING OFF.

⑥ RAISE THE WORK TO A HEIGHT SUFFICIENT TO KEEP THE BLADE FROM STRIKING THE FLOOR. IF THE WORK CANNOT BE RAISED, LIMIT THE STROKE.

Fig. 3-3. Care of saws is important to their long life.

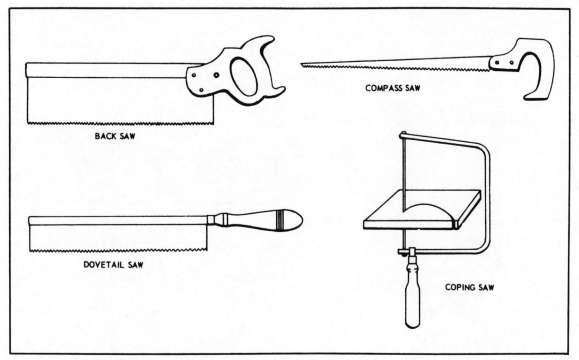

Fig. 3-4. Keyhole or compass saw and related saws.

Fig. 3-5. Backsaw and miter box.

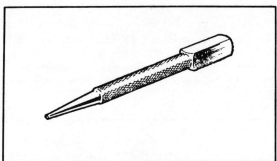

Fig. 3-7. Nail set is used for setting finish nails.

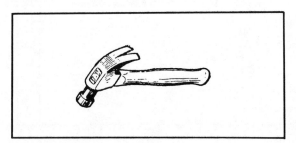

Fig. 3-6. Typical claw hammer.

tion is a good feature.

A *combination square* (Figs. 3-9 and 3-10) has many uses. It can be used to check corners and cuts for squareness, to lay out lines for 45-degree cuts, as a depth guage, and as a bench rule. Good ones have a vial built into the head so it can be used as a level and as a scriber.

A *carpenter's level* (Fig. 3-11) should be at least 24 inches long and contain three vials so it can be

Fig. 3-8. Flex tape rule and other measuring devices.

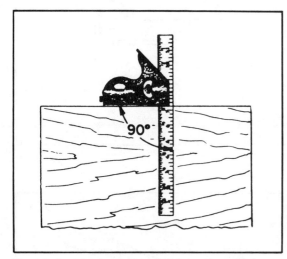

Fig. 3-9. Combination square used for cutting a 90-degree angle.

Fig. 3-10. The combination square is also handy for cutting other angles.

used to check both horizontal and vertical planes. Some have a special vial on one end intended for checking 45-degree angles and slopes—a very useful feature for installing diagonal solid lumber panels.

A *chalk line* (Fig. 3-12) is a good idea because

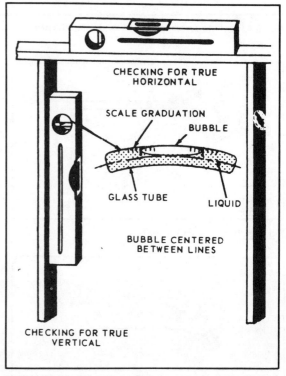

Fig. 3-11. The carpenter's level is useful for trueing doorframes, solid lumber paneling, and other interior members.

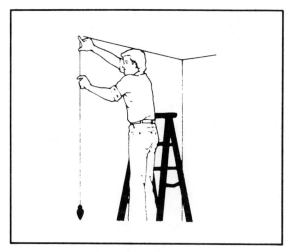

Fig. 3-12. Chalk line being snapped (courtesy Georgia-Pacific).

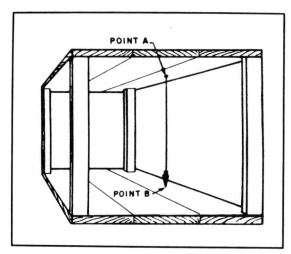

POINT A

POINT B

Fig. 3-13. Some chalk line holders can also be used for plumb bobs.

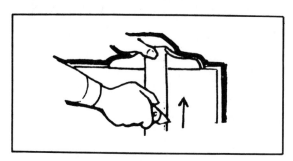

Fig. 3-14. Utility knife (courtesy Georgia-Pacific).

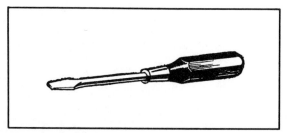

Fig. 3-15. Screwdriver.

it marks long, straight lines. Strong string 50 to 100 feet long is encased with a quantity of chalk dust. When the string is stretched tightly between two points, then snapped, the chalk on the string marks the line. A quality line is refillable with chalk dust and does double duty as a plumb bob (Fig. 3-13), which is useful for making vertical lines.

There are many types of *utility knives* available (Fig. 3-14), but choose one with replaceable, retractable blades. Its uses include making cut lines, scoring gypsum board, and cutting acoustical tile.

Screwdrivers (Fig. 3-15) are best selected in a set. Then you can match the driver to the screw with as much torque as necessary without fear of damage to the tool or the fastener. Many sets include one or two drivers with special heads for Phillips head screws.

Wood chisels (Fig. 3-16) come in a set with sizes ranging from 1/4 inch to at least 1 inch. Good ones cost more, but last indefinitely when treated respectfully. Typical applications are making cutouts in solid lumber panels, shaping notches, and doing planing jobs when there is no room for the plane.

Portable Power Tools

The *electric drill* (Fig. 3-17) is a workhorse because

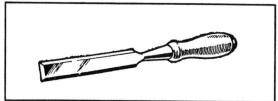

Fig. 3-16. Wood chisel.

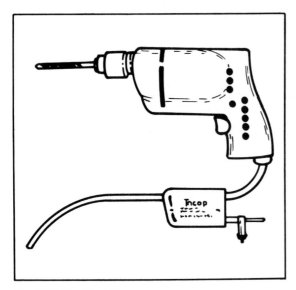

Fig. 3-17. Electric drill.

of the many accessories it will drive. Drilling, sanding, buffing, rotary filing, and screwdriving are a few of the many applications that make it a first choice in power tools. The tool must have the necessary features, however, for the choice to be practical. You will want double insulation to eliminate the need for grounding a 3/8-inch chuck capacity, a trigger switch with lock-on button, variable speed so you can match the rpms to the accessory being used, an adjustable knob to maintain a particular speed, a switch to reverse the chuck's rotation so you can remove screws as well as drive them, a chuck that can be locked with a geared key, and an auxiliary handle usable on either side of the tool's housing when a two-handed grip is advisable.

The portable *circular saw* (Fig. 3-18), often called a cutoff saw, is used for crosscutting, ripping, beveling, mitering, and doing other routine operations much faster than they can be done with a handsaw—an important point when you're cutting a few hundred solid lumber panels. Models range in prices from about $20 to several hundred dollars, but something in the $50 to $75 range should be a reasonable choice for the average home craftsperson who will install more than just solid lumber paneling.

Capacity is often judged in terms of blade diameter. This is a logical point at which to start since the larger the blade, the greater the tool's depth of cut. More important to the average worker, however, is whether the blade will cut through a 2 × 4 at 45 degrees as well as 90 degrees. Most saws will do so whether they have a 6 1/2-inch blade or a 10-inch blade. Generally, the larger the blade, the heavier and the more expensive the tool will be. Replacement blades will also cost more. Why pay for capacity you may never need or tolerate excess weight that can be very tiring?

Do look for double insulate; an easy-to-grip, top-side handle with a built-in trigger switch; at least a 1-hp motor; and an automatic clutch that lets the blade slip should it jam in a cut. All tools have a spring-loaded blade guard that lifts during a cut and returns to cover the blade when the cut is done. Some models have an electronic brake that stops the blade as soon as the trigger is released.

The portable circular saw can be dangerous if used incorrectly. A common negative occurence is called *kickback*—the blade binds and tends to travel backwards so the tool moves toward the operator. To avoid it, cut straight, never force the tool, never extend your reach, keep saw blades sharp, and stand so you are not in line with the tool itself.

There are many types of circular saw blades, and it's wise to have an assortment. The one supplied with the tool will be a combination blade, good for both crosscutting and ripping lumber. For the best cuts on plywood, choose a special plywood blade. A hollow ground will leave smooth edges on both lumber and plywood. There are also special crosscutting blades and rip blades, which should be used when there is much work to be done in either area.

The *saber saw* is excellent for cutting curved lines and for doing internal cutouts, but is also good for crosscutting, ripping, and other routine work. It is very popular for cutting narrow solid lumber panels. The blade moves up and down and is gripped only at one end, which is why the tool can work in the center of a plywood panel without needing a lead-in cut from an edge. There are many types of blades to cut metal, ceramic tile, and other

54

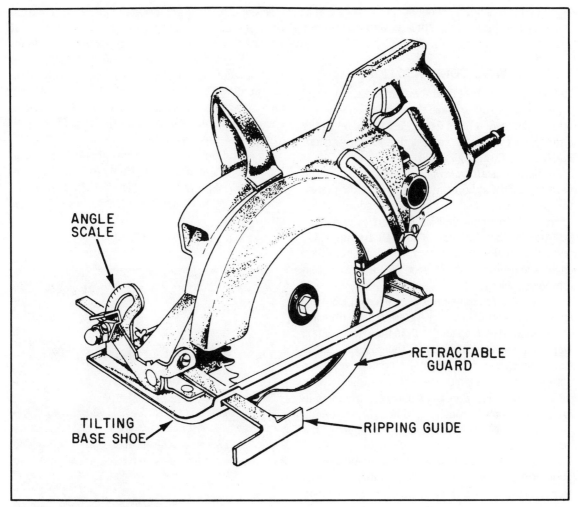

ANGLE
SCALE

RETRACTABLE
GUARD

TILTING
BASE SHOE

RIPPING GUIDE

Fig. 3-18. Portable circular saw.

materials as well as wood. They are interchangeable and inexpensive enough to be considered disposable.

A good saber saw will be double insulated, have trigger switch-controlled variable speed (top speed of about 3200 to 3500 strokes per minute), at least a 1/3- hp motor, the capacity to cut through a 2 × 4, and an adjustable base for bevel cutting.

Depending on the type of solid lumber paneling you decide to install, you may want a *finishing sander*, often called a *pad sander*, so you don't have to spend hours doing smoothing chores. The tool's motor activates a soft pad over which the abrasive

paper is held taut. The pad's action may be reciprocal (to and fro), orbital (moving in tiny circles), or both. The orbital action, will produce the smoothest finishes as long as the orbit is small (1/8 inch or less) and the orbits per minute (opm) is high (9,000 to 10,000 range). The reciprocal action does okay too, but it's best to use a coarse sandpaper when much material must be removed.

Look for double insulation, ball-bearing construction, a direct motor-to-pad drive, a pad in the 3- × 7-inch range, and clamps that will hold the abrasive paper tautly across the pad. The sandpaper must move with the pad. Some tools can be equipped with a lamb's-wool bonnet so they can be used for polishing and buffing as well as sanding.

WALL CONSTRUCTION

Before you install solid lumber paneling, you need to understand and prepare the surface on which it will be mounted. This is usually a wall; so let's take a quick look at home construction and wall framing (Fig. 3-19 through 3-28).

The term *wall framing* includes primarily the vertical studs and horizontal members (sole plates, top plates, and window and door headers) of exterior and interior walls that support ceilings, upper floors, and the roof. The wall framing also serves as a nailing base for wall-covering materials such as gypsum board, plywood paneling, and solid lumber paneling.

The wall framing members used in conventional construction are generally nominal 2-×-4-inch studs spaced 16 inches on center. Depending on the thickness of the covering material, 24-inch spacing might be considered. Top plates and sole plates are also nominal 2 × 4 inches in size. Headers over doors or windows in load-bearing walls consist of doubled 2-×-6-inch and deeper members, depending on span of the openings.

The requirements for wall-framing lumber are good stiffness, good nail-holding ability, freedom from warp, and ease of working. Species used may include Douglas fir, the hemlocks, southern pine, the spruces, pines, and white fir. The grades vary by species, but it is common practice to use the third grade for studs and plates and the second grade for headers over doors and windows.

All framing lumber for walls should be reasonably dry. Material at about 15-percent moisture content is desirable, with the maximum allowable considered to be 19 percent. When material with the higher moisture content is used, it's best to allow the moisture content to reach in-service conditions before applying interior trim.

Ceiling height for the first floor is 8 feet under most conditions. It is common practice to rough-frame the wall (subfloor to top of upper plate) to a height of 8 feet 1 1/2 inches. In platform construction, precut studs are often supplied to a length of 7 feet 8 5/8 inches for a plate thickness of 1 5/8 inches. When dimension material is 1 1/2 inches thick, precut studs will be 7 feet 9 inches long. This height allows the use of 8-foot-high drywall sheets, or six courses of rock lath, and still provides clearance for floor and ceiling finish or for plaster grounds at the floor line.

Second-floor ceiling heights should not be less than 7 feet 6 inches in the clear, except that portion under sloping ceilings. Half of the floor area, however, should have at least 7-foot, 6-inch clearance.

Two general types of wall framing are com-

Fig. 3-19. Home construction begins with finding the property corner.

Fig. 3-20. The foundation is poured.

Fig. 3-21. Lumber is delivered and the floor is begun.

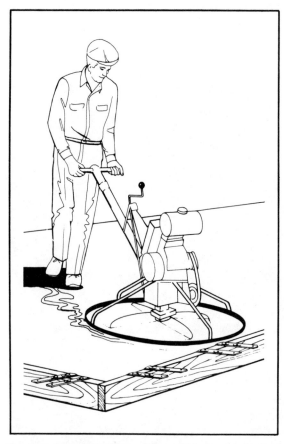

Fig. 3-22. A concrete finisher is brought in for slab flooring or patios (courtesy Simpson Strong-Tie).

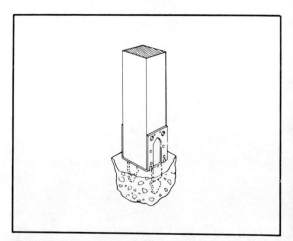

Fig. 3-23. Posts are anchored for porches and patios (courtesy Simpson Strong-Tie).

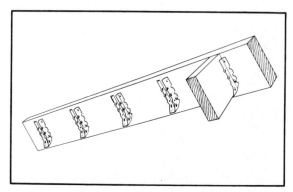

Fig. 3-24. Joist hangers are installed (courtesy Simpson Strong-Tie).

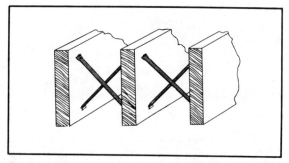

Fig. 3-25. Bridging is installed between floor joists (courtesy Simpson Strong-Tie).

Fig. 3-26. Tongue-and-groove lumber is often used for flooring.

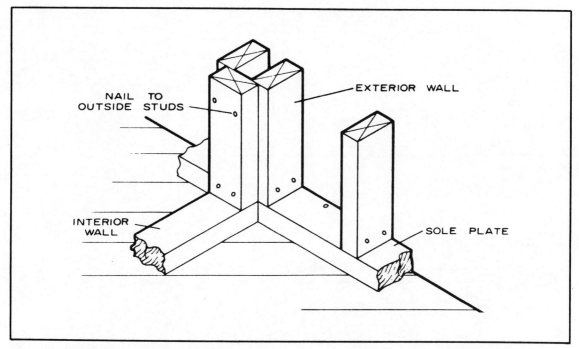

Fig. 3-27. Joining of interior and exterior wall.

monly used—platform construction and balloon-frame construction. The *platform method* (Fig. 3-29) is more often used because of its simplicity. *Balloon framing* (Fig. 3-30) is generally used where stucco or masonry is the exterior covering material in two-story houses. These illustrations will help you understand the components and construction of your own home's walls.

INTERIOR WALLS

The interior walls in a house with conventional joist

Fig. 3-28. Framing walls with post fastener plates (courtesy Simpson Strong-Tie).

and rafter roof construction (Figs. 3-31 through 3-42) are normally located to serve as bearing walls for the ceiling joists, as well as room dividers. Walls located parallel to the direction of the joists are commonly nonload-bearing. Studs are nominal 2 × 4 inches in size for load-bearing walls, but can be 2 × 3 inches in size for nonload-bearing walls. Most contractors use 2 × 4s throughout, however. Spacing of the studs is usually controlled by the thickness of the covering material. For example, 24-inch stud spacing will require 1/2-inch gypsum board for drywall interior covering or equivalent solid lumber paneling.

The interior walls are assembled and erected in the same manner as exterior walls, with a single bottom sole plate and double top plates. The upper top plate is used to tie intersecting and crossing walls to each other. You can also use a single framing stud at each side of a door opening in nonload-bearing walls. Location of the walls and size and spacing of the studs are determined by the room size desired and the type of interior covering selected.

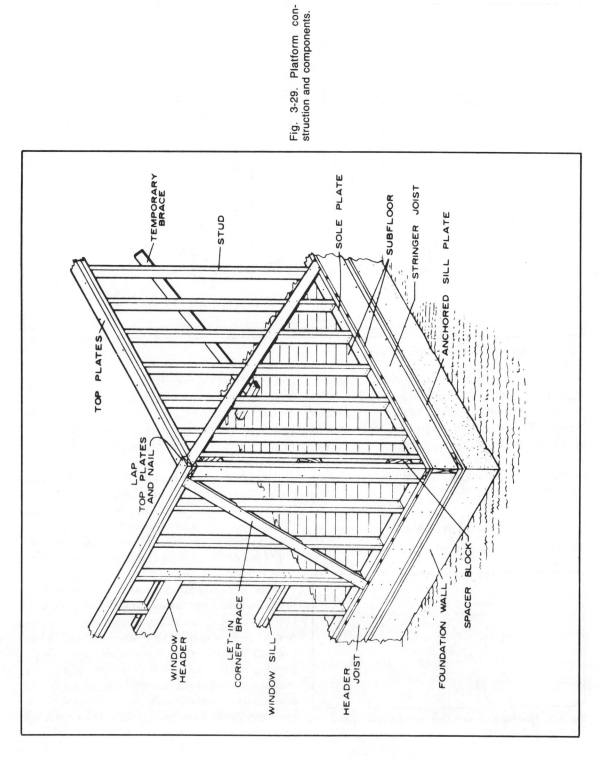

Fig. 3-29. Platform construction and components.

TEMPORARY BRACE

STUD

SOLE PLATE

SUBFLOOR

STRINGER JOIST

ANCHORED SILL PLATE

TOP PLATES

LAP TOP PLATES AND NAIL

WINDOW HEADER

LET-IN CORNER BRACE

WINDOW SILL

HEADER JOIST

FOUNDATION WALL

SPACER BLOCK

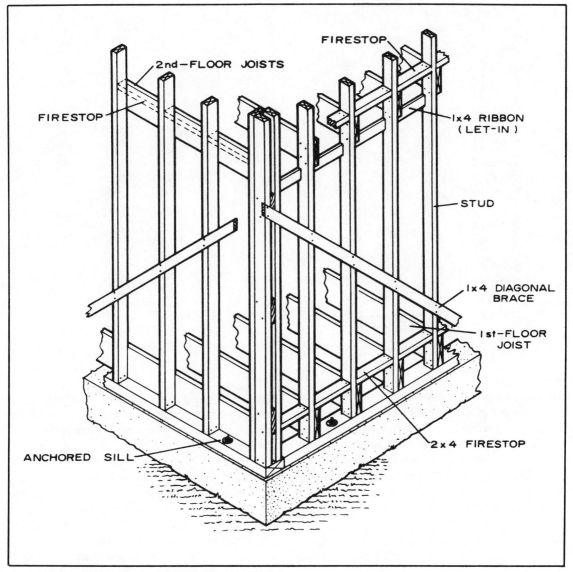

FIRESTOP

2nd—FLOOR JOISTS

FIRESTOP

1x4 RIBBON (LET-IN)

STUD

1x4 DIAGONAL BRACE

1st-FLOOR JOIST

2x4 FIRESTOP

ANCHORED SILL

Fig. 3-30. Balloon framing and components.

PREPARING INSULATION

Since the mid-1970s, consumers in the United States have experienced continuous, sometimes sharp, increases in the cost of energy used to heat their homes. Many homeowners have taken low-cost energy conservation steps, including increasing the levels of insulation in the home. This is an especially practical step when you are installing or remodeling walls within your home (Fig. 3-43). A poorly insulated, gypsum-covered wall can be disassembled, insulated, and covered with solid lumber paneling to add both beauty and function to the home. Depending on which part of the home is insulated, this investment may help you pay for your paneling project in just a few years.

Understanding how heat moves will make it

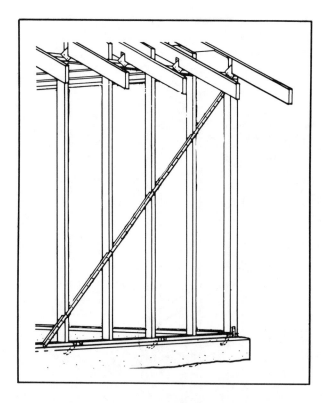

Fig. 3-31. Wall bracing detail with straps (courtesy Simpson Strong-Tie).

easier to understand the benefits of adding insulation. Heat moves in three ways: *conduction* (by contact), *convection* (by a carrying fluid such as air or water), and *radiation* (by electromagnetic waves). When heat is lost from a house, all three of these processes are in action. The measure of the movement of heat through a given material if often called its *conductance*, but *thermal transmission* may be a less confusing term since heat can move through some materials by radiation and convection as well as by conduction. Thermal transmission, or *U value*,

is a measure of how many units of heat (Btus) pass through a given thickness of a square foot of the material in an hour at a certain temperature difference.

The ability of a material to resist heat flow, or to insulate, is indicated by its *R value*. The higher the R value of a material, the better an insulator it is. The R value is the inverse of the U value; that is, $R = 1/U$ and $U = 1/R$. It is important to know this distinction because while insulations are compared by R values, some charts may list only the U value of a particular building material. When different materials are combined, as in the wall of a home, all the U values must be changed to R values before the total R value can be determined. Although R values can be added to get a total, U values cannot.

It is also helpful to note that heat always attempts to reach equilibrium. This means that heat always moves toward cold and will take the path of least resistance. How fast heat will travel from one area to another is directly related to the dif-

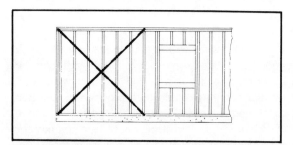

Fig. 3-32. Wall cross bracing with straps (courtesy Simpson Strong-Tie).

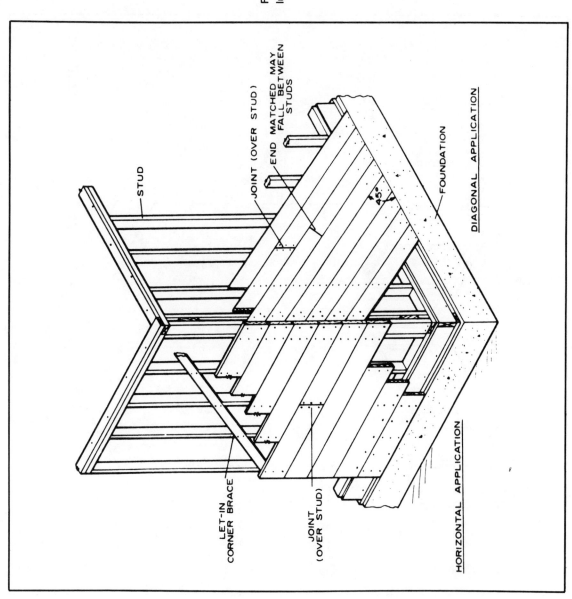

Fig. 3-33. Exterior solid lumber paneling details.

STUD

JOINT (OVER STUD)

END MATCHED MAY
FALL BETWEEN
STUDS

FOUNDATION

45°

DIAGONAL APPLICATION

LET-IN
CORNER BRACE

JOINT
(OVER STUD)

HORIZONTAL APPLICATION

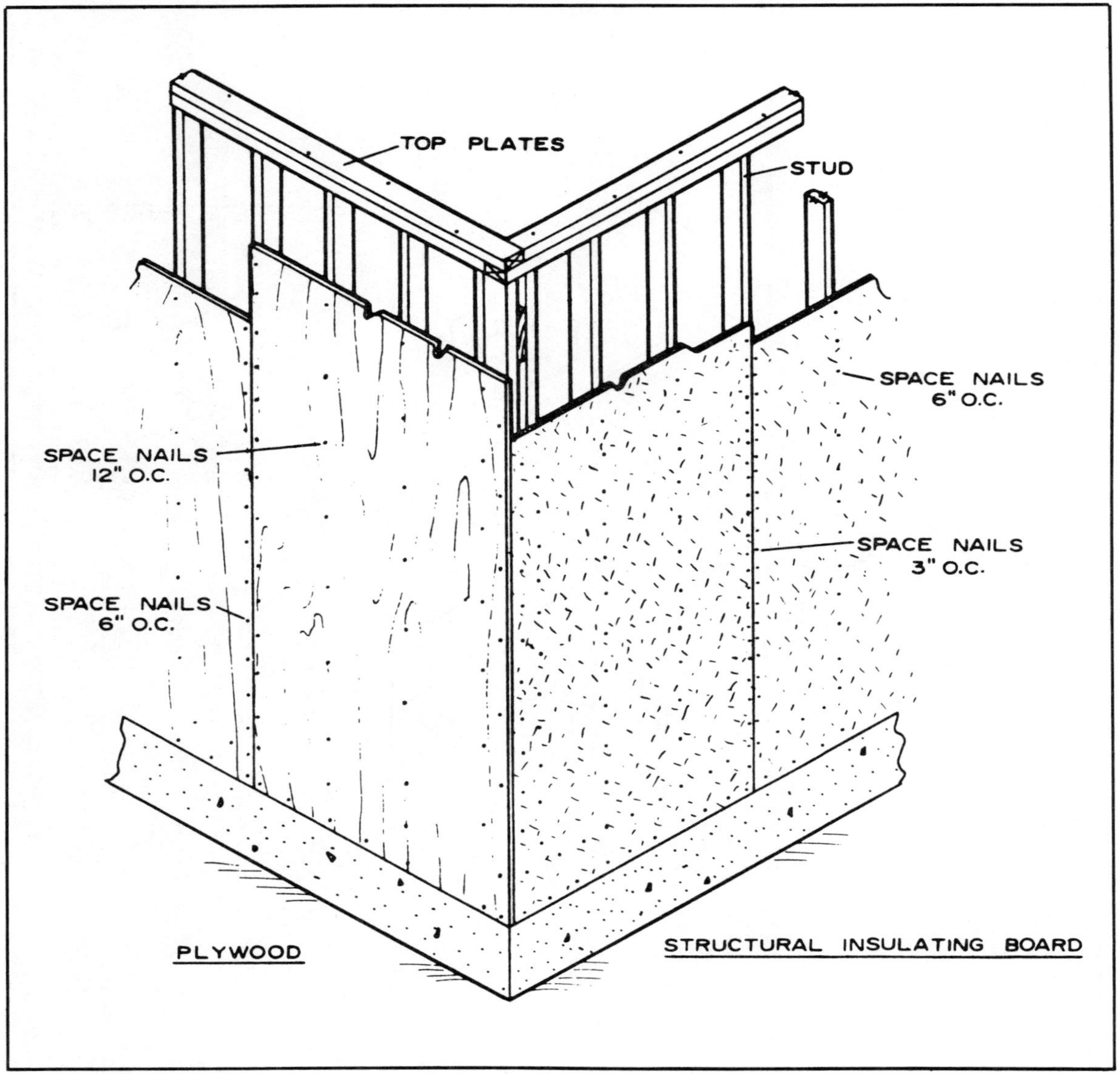

Fig. 3-34. Exterior plywood details.

STUD

TOP PLATES

SPACE NAILS 6" O.C.

SPACE NAILS 3" O.C.

SPACE NAILS 12" O.C.

SPACE NAILS 6" O.C.

STRUCTURAL INSULATING BOARD

PLYWOOD

Fig. 3-35. The first wall goes up.

Fig. 3-36. Attaching the wall to subflooring.

Fig. 3-37. Framing the second wall on the floor.

Fig. 3-38. Plates are marked and studs installed for wall.

66

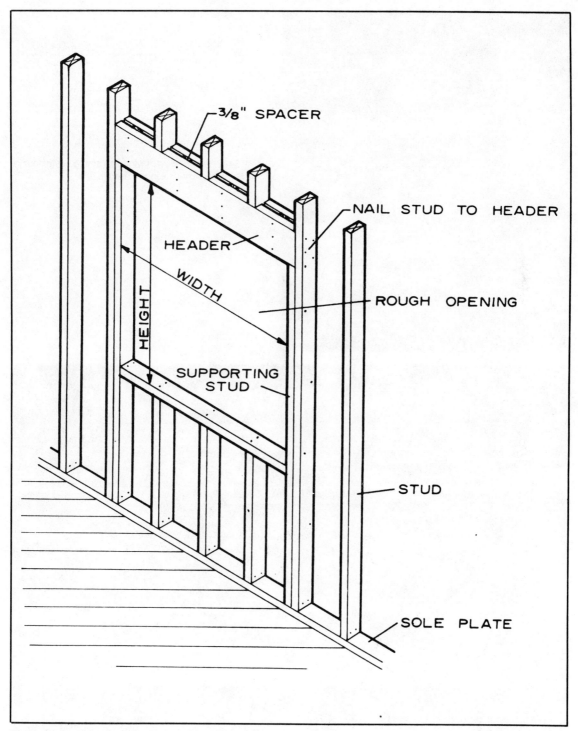

Fig. 3-39. Stud details for typical window construction.

Fig. 3-40. Another view of window framing. Note double studs at outside of window frame.

Fig. 3-41. Window frame and header.

Fig. 3-42. Side wall attached to end wall and braced in position.

ference in temperature between the two areas. For example, twice as much heat will be lost from a 70-degrees Fahrenheit house on a day when the outside temperature averages 0 degrees Fahrenheit than on a day averaging 35 degrees Fahrenheit outside. Because the temperature difference is twice as great, heat escapes twice as fast.

INSULATION MATERIALS

Most insulations get their R value by creating millions of dead air spaces to slow the conduction of heat. Some insulations attempt to reduce the radiant heal losses by using reflective material.

Commercial insulation is manufactured in a variety of forms and types, each with advantages for specific uses. Materials commonly used for insulation may be grouped in the following general classes:

☐ Flexible insulation (blanket and batt)
☐ Loose-fill insulation
☐ Reflective insulation
☐ Rigid insulation (structural and non-structural)
☐ Miscellaneous types

Flexible insulation is made in two types: blanket and batt. *Blanket insulation* (Fig. 3-44) is furnished in rolls or packages in widths suited to 16- and 24-inch stud and joist spacing. Usual thicknesses are 1 1/2, 2, and 3 inches. The body of the blanket is made of felted mats of mineral or vegetable fibers, such as rock or glass wool, wood fiber, and cotton. Organic insulations are treated to make them resistant to fire, decay, insects, and vermin.

Most blanket insulation is covered with paper or other sheet material with tabs on the side to fasten to studs or joists. One covering sheet serves as a vapor barrier to resist movement of water vapor and should always face the warm side of the wall. Aluminum foil or asphalt or plastic-laminated paper are commonly used as barrier materials.

Batt insulation (Fig. 3-45) is also made of fibrous material to preformed thicknesses of 4 and 6 inches for 16- and 24- inch joist spacing. It is supplied with or without a vapor barrier. One friction type of fibrous glass batt is supplied without a covering and is designed to remain in place without the normal fastening methods.

Loose-fill insulation (Fig. 3-46) is usually composed of materials used in bulk form, supplied in bags or bales, and placed by pouring, blowing, or packing by hand. This type of insulation includes rock or glass wool, wood fibers, shredded redwood bark, cork, wood pulp products, vermiculite, sawdust, and shavings.

Fill insulation is suited for use between first-floor ceiling joists in unheated attics. It is also used in side walls of existing houses that were not insulated during construction. Holes can be punched into the existing wall and loose-fill insulation added between studs. The wall can then be covered with solid lumber paneling.

Reflective insulation (Fig. 3-47) of the foil type

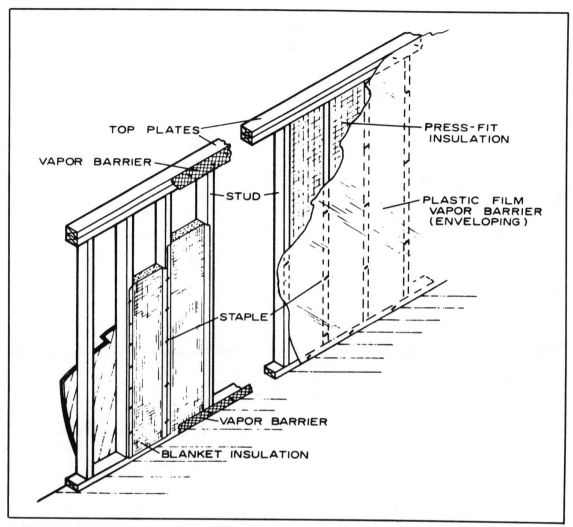

Fig. 3-43. Insulating with blanket (left) and press-fit (right) insulation.

Fig. 3-44. Blanket insulation.

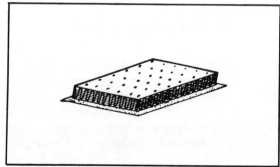

Fig. 3-45. Batt insulation.

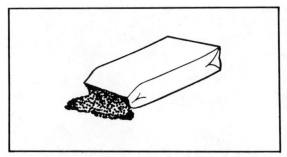

Fig. 3-46. Fill insulation.

Fig. 3-48. Rigid insulation.

is sometimes applied to blankets and to the stud-surface side of drywall or paneling. Metal foil suitably mounted on some supporting base makes an excellent vapor barrier. The type of reflective insulation shown in Fig. 3-47 includes reflective surfaces and air spaces between the outer sheets.

Rigid insulation is usually a fiberboard material manufactured in sheet and other forms (Fig. 3-48). Rigid insulations are also made from such materials as inorganic fiber and glass fiber, although these are not commonly used in a house in this form. The most common types are made from processed wood, sugar cane, or other vegetable products. Structural insulating boards, in densities ranging from 15 to 31 pounds per cubic foot, are fabricated in such forms as building boards, roof decking, sheathing, and wallboard. While they have moderately good insulating properties, their primary purpose is structural (Fig. 3-49 and 3-50).

Some insulations don't fit in the classifications just discussed, such as insulation blankets made up of multiple layers of corrugated paper. Other types, such as lightweight vermiculite and perlite aggregates, are sometimes used in plaster as a means of reducing heat transmission. Expanded

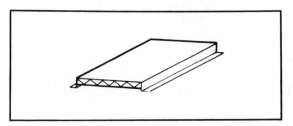

Fig. 3-47. Reflective insulation.

polystyrene and urethane plastic foams may be molded or foamed in place. Urethane insulation may also be applied by spraying. Polystyrene and urethane in board form can be obtained in thicknesses from 1/2 to 2 inches.

INSULATING WALLS

The additional equipment and labor required to add insulation to existing walls makes it the lowest priority in most homes. The important exception to this rule is when remodeling plans make adding wall insulation cheap.

If the wall is to be paneled, for example, rigid foam board or flexible insulation can be applied before the new paneling. When remodeling involves opening the wall cavity, it is quite simple to put in the properly sized of fiberglass blanket. Even when a paper or foil-faced insulation is used, it is smart to install a polythylene vapor barrier over the insulation. The long, fairly penetrable cracks on each side of the blankets are then sealed. Unfaced insulation can also be used, and may be cheaper.

Without tearing the wall apart, about the only way to insulate walls is to blow in the insulation. The most common materials are cellulose, fiberglass, and rock wool. This technique usually requires professional installation, which is part of the reason the cost is higher. Cellulose manufacturers claim the longer fibers of the mineral wools can snag on nail points or splinters, block the cavity, and leave voids. This is not a problem with cellulose because it is a finer particle. The other disadvantage to mineral wools is that the access hole must be larger.

Fig. 3-49. Structural insulating boards ready for installation.

Fig. 3-50. Framing of home with structural insulating boards in place.

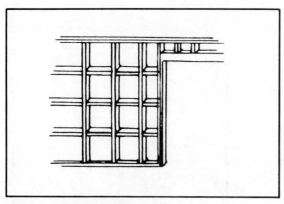

Fig. 3-51. Verticals and horizontals.

In either case, a two-hole method is recommended for good coverage. In each story two holes, one high and one low, are drilled in each cavity or any space between two studs. The low hole should be no more than 4 feet from the bottom plate because cellulose will fill adequately downward that far. The upper hole should be no more than 18 inches from the top plate, because that is the distance above the hole that will fill adequately. If there is more than 4 feet between holes, another hole should be drilled and blown. Before the cavity is filled, a plumb line should be dropped through the hole to check for fire blocking. If the line goes slack before the bottom plate is reached, then there is blocking and another hole may have to be drilled. The extra framing around wall openings may also call for additional holes to be drilled.

Some other checks must be made before the insulation is blown. The perimeter of the attic and basement must be inspected to make sure the cavities are closed at these points. Otherwise, cellulose will start filling the basement or attic while the wall is being blown. If the blower doesn't cut off in the appropriate time, it could be an indication that some additional area is being filled. Other areas where the wall might not be sealed are underneath sink cabinets, under closed-in stairways, and behind the soffits above wall-hung cabinets. Any poorly sealed penetrations through the wall would also cause problems. The perimeter check should pick these out and locate any heating ducts which run up the exterior wall.

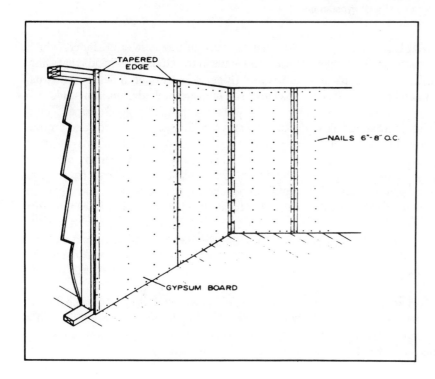

Fig. 3-52. Drywall construction using vertical gypsum board.

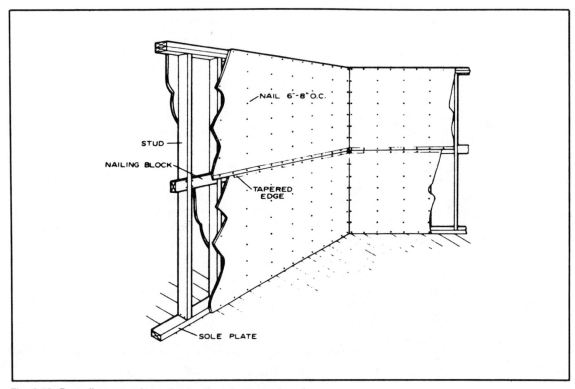

Fig. 3-53. Drywall construction using horizontal gypsum board.

PREPARING WALLS

Now it's time to get the walls ready for the installation of solid lumber paneling. The same steps can be modified to prepare ceilings and other surfaces for solid lumber paneling.

We'll concern ourselves with decorating an existing room, since this is the most common project. If you're working on new construction, the problems are fewer. In paneling a new room, you are building out from straight, true walls. In redecorating, particularly in an older home, you'll have the challenge of innovating.

First, let's review typical room construction. Walls are usually constructed of 2 × 4 studs (verticals) with plates (horizontals) and floor and ceiling (Fig. 3-51). Studs are usually set every 16 inches and at doors and windows. Very often you'll find double construction (two 2 × 4s) at doorways and windows. To locate each stud, tap the finished wall with a hammer. At a stud the sound will be solid.

You can also use a magnetic stud finder, which locates the nails in the studs. Nail heads that show in drywall (Figs. 3-52 through 3-54), baseboards, or chair rails indicate nails driven into studs.

Most walls are in pretty good shape. You'll want to remove moldings from the top and bottom of walls next. Also, if there is loose plaster, tear it out and build out the wall with furring or repair it with patching plaster. It's best to remove moldings around doors and windows also. Do so carefully to avoid splitting, with a wedge or a carpenter's pry bar. You can also drive the nails clear through moldings with hammer and nail set.

You may need to build out with studding and/or shims in certain places to match the vertical plane of the rest of the wall. This may be necessary if you discover a void in the wall. It must also be done if you remove an electric outlet or are filling in a larger opening (such as to change a window into a smaller pass-through.

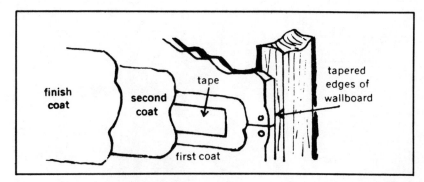

Fig. 3-54. Taping wallboard.

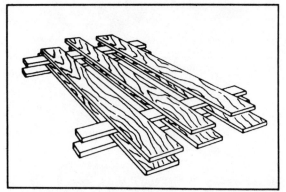

Fig. 3-55. Stacking solid lumber paneling with spacers (courtesy Champion International).

PREPARING WOOD

To minimize the amount of expansion and contraction that could occur during the changes in moisture content, remove individual pieces of solid lumber paneling from the package and allow them to acclimate to room conditions for at least 48 hours before they are installed. Stack the panels with lumber or spacers between layers and space between individual pieces (Fig. 3-55). If high humidity exists, allow a longer period for preconditioning.

FURRING

You may have to fur some walls to make them true.

Fig. 3-56. Leveling furring strips (courtesy Georgia-Pacific).

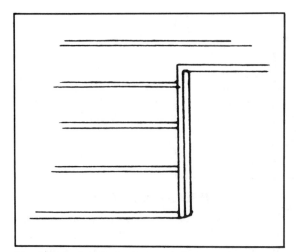

Fig. 3-57. Laying furring to an opening.

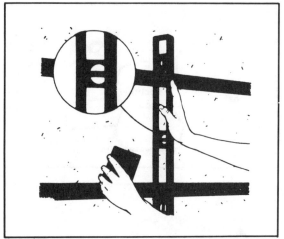

Fig. 3-59. Checking for level (courtesy Georgia-Pacific).

First, paint the ceiling and baseboards, molding, trim, and window and door casings that will remain with your new paneling. Remove those that will be eliminated or replaced. If your walls are in good condition, you can glue and nail solid lumber paneling directly into studs, butted neatly against existing trim. On uneven, badly cracked, or very rough walls, solid lumber paneling can be nailed to a simple framework of 1-×-4-inch or 1-×3-inch furring strips (Figs. 3-56 through 3-60) of any wood species and grade that is kiln-dried. Fitting shims (ordinary wood shingles) under the strips will even up severe surface gaps.

To figure furring strip requirements for vertical paneling, measure wall width to determine the length of one horizontal strip and wall height to figure how many strips you'll need, spaced about 24 inches on center. Furring strips must also be nailed around windows, doors, and all other openings; so add these measurements to the total. At windows and doors, nail strips flush with inside or wall edges of jambs so that build-up strips can be added later.

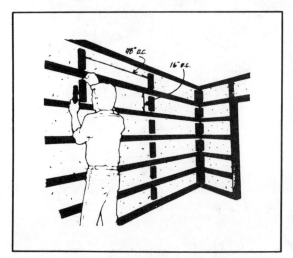

Fig. 3-58. Installing blocks between furring strips (courtesy Georgia-Pacific).

Fig. 3-60. Shimming furring strips (courtesy Georgia-Pacific).

GETTING READY

You're just about ready to install your solid lumber paneling. You've selected the tools; prepared, insulated, and furred the wall; and prepared the panels.

Before you begin the actual installation of solid lumber panels, however, you'll need to consider how the panels will be fastened to the walls and other surfaces. In Chapter 4, we'll look at solid lumber paneling fasteners and adhesives.

Chapter 4

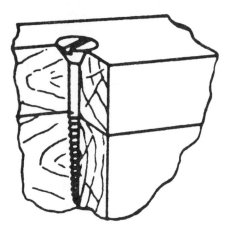

Fasteners and Adhesives

WOULDN'T IT BE GREAT IF YOU COULD JUST hold your solid lumber panel up against the wall for a second, step back, and see it stay up? Then, when you want to change the design or replace a piece, you simply grasp it and lift it off.

Unfortunately, until the invention of "antigravity solid lumber paneling," you'll have to use fasteners and adhesives to install your strip paneling on walls, ceilings, and other surfaces (Figs. 4-1 through 4-3). That's what this chapter is about.

The fastening devices most commonly used are usually made of metal. They are classified as: nails, screws, bolts, driftpins, corrugated fasteners, and timber connectors. Not all of these are used in fastening solid lumber paneling to walls, however. Let's consider those that are.

NAILS

The standard nail for installing a variety of wood products is the *wire nail*, so called because it is made from steel wire. There are many types of nails, all of which are classified according to use

and form. The wire nail is round-shafted, straight, and pointed, and it varies in size, weight, size and shape of head, type of point, and finish. There is a nail for all normal requirements of construction and framing.

There are a few rules to follow when using nails in building. A nail, whatever the type, should be at least three times as long as the thickness of the wood it is intended to hold. Two-thirds of the nail's length is driven into the second piece for proper anchorage, while 1/3 anchors the piece being fastened. Nails should be driven at an angle slightly toward each other and should be carefully placed to provide the greatest holding power. Nails driven with the grain don't hold as well as nails driven *across* the grain. A few nails of proper type and size, properly placed and driven, will hold better than a great many driven close together.

Nails are generally the cheapest and easiest fasteners to apply. In terms of holding power alone, nails provide the least amount; screws of comparable size provide more, and bolts provide the

78

Fig. 4-1. Special fasteners and adhesives can make difficult installations easy (courtesy California Redwood Association).

Fig. 4-2. Fasteners can be installed so that they don't show (courtesy California Redwood Association).

Fig. 4-3. Unseen fasteners allow the beauty of the solid lumber paneling to be used in a variety of patterns (courtesy California Redwood Association).

greatest amount. Although screws and bolts can be used for installing solid lumber paneling, nails are more common fasteners.

Common wire nails and *box nails* (Fig. 4-4) are

the same except that the wire sizes are one or two numbers smaller for a given length of the box nail. The common wire nail is used for construction of housing framing. The common wire nail and the

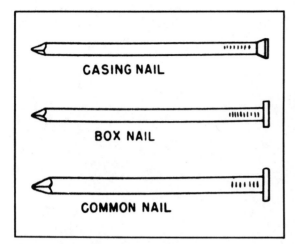

Fig. 4-4. Common wire nails and box nails.

box nail are generally used for structural construction.

The *finishing nail* (Fig. 4-5) is made from finer wire and has a smaller head than the common nail. It may be set below the surface of the wood into which it is driven and will leave only a small hole that is easily puttied up. It is generally used for interior or exterior finishing work and for finished carpentry and cabinetmaking. The finishing nail is the most common type of nail used in the installation of solid lumber paneling because it offers good holding power while being unseen when set below the surface of the wood.

The *duplex nail* (Fig. 4-6) is made with what may appear to be two heads. The lower head, or *shoulder,* is provided so that the nail may be driven securely to give maximum holding power. The upper head projects above the surface of the wood to make its withdrawal simple. The reason for this

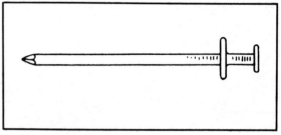

Fig. 4-6. Duplex head nail.

design is that the duplex nail is not meant to be permanent. It is used in the construction of temporary structures, such as scaffolding and staging, and is classified for temporary construction. You can use a duplex nail as you begin solid lumber panel placement so that the initial pattern board is removable. You can also use the duplex nail for constructing scaffolding for the installation of solid lumber paneling on a high wall.

Roofing nails (Fig. 4-7) are round-shafted, diamond-pointed, galvanized nails of relatively short length and comparatively large heads. They are designed for fastening flexible roofing materials and for resisting continuous exposure to weather. Roofing nails are rarely used for installing solid lumber paneling.

SIZING NAILS

Nail sizes are designated by the term *penny,* abbreviated *d* (Fig. 4-8). This term designates the length of the nail—1d, 2d, etc.—which is the same for all types of nails. The approximate number of nails per pound varies according to the type and size. The wire gauge number varies according to type.

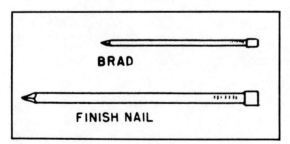

Fig. 4-5. Finish nails and brads.

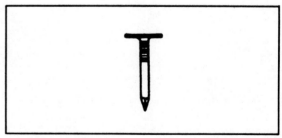

Fig. 4-7. Roofing nail.

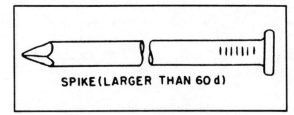

Fig. 4-8. A 60d or 60-penny nail.

Figure 4-9 shows the relative size of common wire nails. The related Fig. 4-10 offers information on the length, wire gauge, and approximate number of nails to the pound. To further help you in selecting nails for your solid lumber paneling project, Table 4-1 offers the size, type and use of nails in all types of rough and finish carpentry, from building cabinets to building bridges.

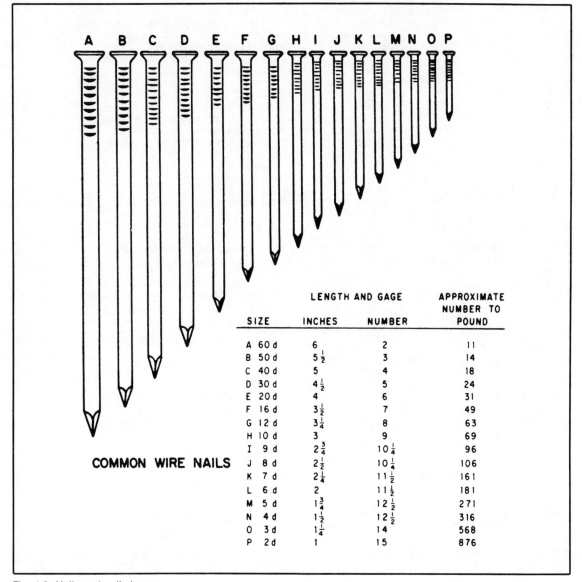

COMMON WIRE NAILS

SIZE	LENGTH AND GAGE		APPROXIMATE NUMBER TO POUND
	INCHES	NUMBER	
A 60 d	6	2	11
B 50 d	$5\frac{1}{2}$	3	14
C 40 d	5	4	18
D 30 d	$4\frac{1}{2}$	5	24
E 20 d	4	6	31
F 16 d	$3\frac{1}{2}$	7	49
G 12 d	$3\frac{1}{4}$	8	63
H 10 d	3	9	69
I 9 d	$2\frac{3}{4}$	$10\frac{1}{4}$	96
J 8 d	$2\frac{1}{2}$	$10\frac{1}{4}$	106
K 7 d	$2\frac{1}{4}$	$11\frac{1}{2}$	161
L 6 d	2	$11\frac{1}{2}$	181
M 5 d	$1\frac{3}{4}$	$12\frac{1}{2}$	271
N 4 d	$1\frac{1}{2}$	$12\frac{1}{2}$	316
O 3 d	$1\frac{1}{4}$	14	568
P 2 d	1	15	876

Fig. 4-9. Nails and nail sizes.

Table 4-1. Nail Uses.

Size	Lgth (in.)[1]	Diam (in.)	Remarks	Where Used
2d	1	.072	Small head	Finish work, shop work.
2d	1	.072	Large flathead	Small timber, wood shingles, lathes.
3d	1 1/4	.08	Small head	Finish work, shop work.
3d	1 1/4	.08	Large flathead	Small timber, wood shingles, lathes.
4d	1 1/2	.098	Small head	Finish work, shop work.
4d	1 1/2	.098	Large flathead	Small timber, lathes, shop work.
5d	1 3/4	.098	Small head	Finish work, shop work.
5d	1 3/4	.098	Large flathead	Small timber, lathes, shop work.
6d	2	.113	Small head	Finish work, casing, stops, etc., shop work.
6d	2	.113	Large flathead	Small timber, siding, sheathing, etc., shop work.
7d	2 1/4	.113	Small head	Casing, base, ceiling, stops, etc.
7d	2 1/4	.113	Large flathead	Sheathing, siding, subflooring, light framing.
8d	2 1/2	.131	Small head	Casing, base, ceiling, wainscot, etc., shop work.
8d	2 1/2	.131	Large flathead	Sheathing, siding, subflooring, light framing, shop work.
8d	1 1/4	.131	Extra-large flathead	Roll roofing, composition shingles.
9d	2 3/4	.131	Small head	Casing, base, ceiling, etc.
9d	2 3/4	.131	Large flathead	Sheathing, siding, subflooring, framing, shop work.
10d	3	.148	Small head	Casing, base, ceiling, etc., shop work.
10d	3	.148	Large flathead	Sheathing, siding, subflooring, framing, shop work.
12d	3 1/4	.148	Large flathead	Sheathing, subflooring, framing.
16d	3 1/2	.162	Large flathead	Framing, bridges, etc.
20d	4	.192	Large flathead	Framing, bridges, etc.
30d	4 1/2	.207	Large flathead	Heavy framing, bridges, etc.
40d	5	.225	Large flathead	Heavy framing, bridges, etc.
50d	5 1/2	.244	Large flathead	Extra-heavy framing, bridges, etc.
60d	6	.262	Large flathead	Extra-heavy framing, bridges, etc.

[1]This chart applies to wire nails, although it may be used to determine the length of cut nails.

SCREWS

The use of screws rather than nails as fasteners may be dictated by a number of factors, including the type of material to be fastened, the need for greater holding power than could be obtained by the use of nails, the finished appearance desired, and the use of a limited number of fasteners. Because solid lumber panels are finish rather than structural members and to achieve the desired finished appearance, screws are normally not used in installing solid lumber panels. They are, however, used for framing, some trim, and many related tasks.

The use of screws rather than nails is more expensive in terms of time and money, but is often necessary to meet requirements for superior results. The main advantages of screws are that they provide more holding power, can be easily tightened to draw the items being fastened securely together, are neater in appearance if properly driven, and may be withdrawn without damaging the material.

The common wood screws (Fig. 4-10) is usually made of unhardened steel, stainless steel, aluminum, or brass. The steel may be bright finished or blued, or zinc-, cadmium-, or chrome-plated. Wood screws are threaded from a gimlet point for approximately 2/3 of the length of the screw. They have a slotted head designed to be driven by an inserted driver.

Wood screws, as shown in Fig. 4-11, are designated according to head size. The most common types are: flathead, ovalhead, and roundhead, both in slotted and Phillips heads (Fig. 4-12).

To prepare wood for receiving the screws, bore a pilot hole the diameter of the screw in the wood to be fastened (Fig. 4-13). Then bore a smaller, starter hole in the piece of wood that is to act as

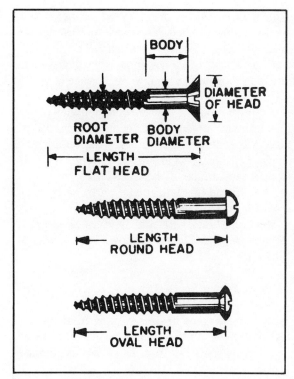

Fig. 4-10. Common screw terms.

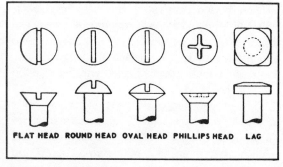

Fig. 4-12. Screw heads.

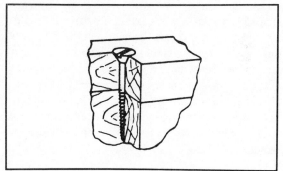

Fig. 4-13. Countersinking the wood screw.

anchor or hold the threads of the screw. The starter hole is drilled with a diameter less than that of the screw threads and to a depth 1/2 or 2/3 the length of the threads to be anchored.

The purpose of this careful preparation is to ensure accuracy in the placement of the screws, to reduce the possibility of splitting the wood, and to reduce the time and effort required to drive the screw. Properly set slotted and Phillips flathead (Fig. 4-14) and ovalhead screws are countersunk

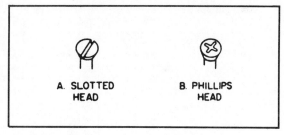

Fig. 4-14. Slotted and Phillips heads.

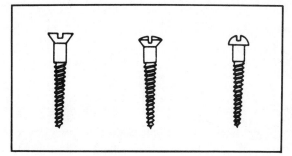

Fig. 4-11. Wood screws.

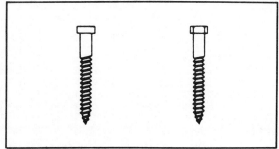

Fig. 4-15. Lag screws.

sufficiently to permit you to cover the head. Slotted roundhead and Phillips roundhead screws are not countersunk, but are driven so that the head is firmly flush with the surface of the wood. The slot of the roundhead screw is left parallel with the grain of the wood.

The proper name for lag screws (Fig 4-15) is *lag bolt, wood screw type*. These screws are often required in construction building. They are longer and much heavier than the common wood screw and have coarser threads which extend from a cone or gimlet point slightly more than half the length of the screw. Lag screws with square- or hexagon-shaped heads are always externally driven, usually by means of a wrench. They are used when ordinary wood screws would be too short or too light and spikes would not be strong enough. For sizes of lag screws, see Table 4-2. Combined with expansion anchors, lag screws are used to frame timbers to existing masonry. Use them to install a wood wall over masonry before covering it with solid lumber paneling.

Expansion shields, or expansion anchors as they are sometimes called, are used for inserting a predrilled hole, usually in masonry. They provide a gripping base, or anchor, for a screw, bolt, or nail intended to fasten an item to the surface in which the hole is bored. The shield can be obtained separately or include the screw, bolt, or nail. After the expansion shield is inserted into the predrilled hole, the fastener is driven into the hole in the shield, expanding the shield and wedging it firmly against the surface of the hole.

For the assembly of metal parts, *sheet metal screws* are used. These screws are made regularly in steel and brass with four types of heads: flat, round, oval, and fillister, as shown in that order in Fig. 4-16.

SIZING WOOD SCREWS

Wood screws come in sizes varying from 1/4 to 6 inches. Screws up to 1 inch in length increase by eighth inches; screws from 1 to 3 inches increase

Table 4-2. Screw Sizes and Dimensions.

Length (in.)	Size numbers																					
	0	1	2	3	4	5	6	7	8	9	10	11	12	13	14	15	16	17	18	20	22	24
1/4	x	x	x	x																		
3/8	x	x	x	x	x	x	x	x	x	x												
1/2		x	x	x	x	x	x	x	x	x	x	x	x									
5/8		x	x	x	x	x	x	x	x	x	x	x	x		x							
3/4			x	x	x	x	x	x	x	x	x	x	x		x		x					
7/8			x	x	x	x	x	x	x	x	x	x	x		x		x					
1				x	x	x	x	x	x	x	x	x	x		x		x		x	x		
1 1/4					x	x	x	x	x	x	x	x	x		x		x		x	x		x
1 1/2					x	x	x	x	x	x	x	x	x		x		x		x	x		x
1 3/4						x	x	x	x	x	x	x	x		x		x		x	x		x
2						x	x	x	x	x	x	x	x		x		x		x	x		x
2 1/4						x	x	x	x	x	x	x	x		x		x		x	x		x
2 1/2						x	x	x	x	x	x	x	x		x		x		x	x		x
2 3/4							x	x	x	x	x	x	x		x		x		x	x		x
3							x	x	x	x	x	x	x		x		x		x	x		x
3 1/2									x	x	x	x	x		x		x		x	x		x
4									x	x	x	x	x		x		x		x	x		x
4 1/2													x		x		x		x	x		x
5													x		x		x		x	x		x
6															x		x		x	x		x
Threads per inch	32	28	26	24	22	20	18	16	15	14	13	12	11		10		9		8	8		7
Diameter of screw (in.)	.060	.073	.086	.099	.112	.125	.138	.151	.164	.177	.190	.203	.216		.242		.268		.294	.320		.372

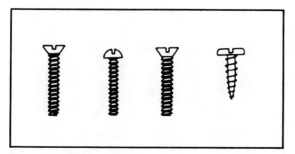

Fig. 4-16. Metal screws.

by quarter inches; and screws from 3 to 6 inches increase by half inches. Screws vary in length and size of shaft. Each length is made in a number of shaft sizes specified by an arbitrary number which represents no particular measurement but indicates relative differences in the diameter of the screws.

Proper nomenclature of a screw includes the type, material, finish, length, and screw size number. The screw size number indicates the wire gauge of the body, drill or bit size for the body hole, and drill or bit size for the starter hole. Table 4-3 provides size, length, gauge, and applicable drill and auger bit sizes for screws.

BOLTS

Although bolts are not a primary fastener in panel installation, they are often used in related construction when great strength is required or when the work must be frequently disassembled. Bolts usually employ nuts for fastening and sometimes washers to protect the surface of the material being fastened. There are four general types of bolts: carriage bolts, machine bolts, stove bolts, and expansion bolts.

Carriage bolts (Fig. 4-17) fall into three categories: square neck or common, finned neck, and ribbed neck. These bolts have round heads that are

not designed to be driven. They are threaded only part of the way up the shaft; usually the threads are two to four times the diameter of the bolt in length. In each type of carriage bolt, the upper part of the shank, immediately below the head, is designed to grip the material in which the bolt is inserted and keep the bolt from turning when a nut is tightened down on it or removed. The finned type has two or more fins extending from the head to the shank. The ribbed type has longitudinal ribs, splines, or serrations on all or part of a shoulder located immediately beneath the head.

Holes that receive carriage bolts are bored to a tight fit for the body of the bolt and counterbored to permit the head of the bolt to fit flush with or below the surface of the material being fastened. The bolt is then driven through the hole with a hammer.

Carriage bolts are chiefly for wood-to-wood application, but may also be used for wood-to-metal applications. If used for wood-to-metal installations, the head should be fitted to the wood item. Metal surfaces are sometimes predrilled and countersunk to permit the use of carriage bolts and metal to metal.

Carriage bolts can be obtained from 1/4 to 1 inch in diameter and from 3/4 to 20 inches long. A common flat washer should be used with carriage bolts between the nut and the wood surface.

Machine bolts (Fig. 4-18) are generally for metal-to-metal installations where close tolerance is important. Selection of the proper machine bolt is made on the basis of head style, length, diameter, number of threads per inch, and coarseness of the thread.

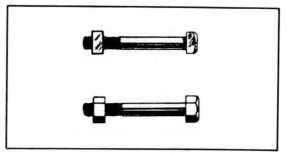

Fig. 4-18. Machine bolts and nuts.

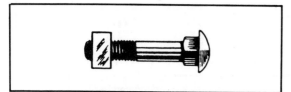

Fig. 4-17. Carriage bolt and nut.

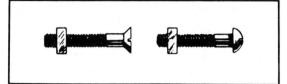

Fig. 4-19. Stove bolts and nuts.

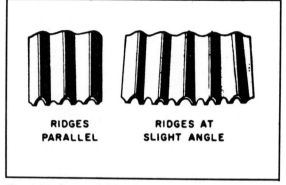

RIDGES PARALLEL   RIDGES AT SLIGHT ANGLE

Fig. 4-21. Corrugated fasteners.

Stove bolts (Fig. 4-19) are less precisely made then machine bolts and are normally used with square nuts and applied metal to metal, wood to wood, or wood to metal. If flatheaded, they are countersunk; if roundheaded, they are drawn flush to the surface.

An *expansion bolt* (Fig. 4-20) is used with an expansion shield to anchor in substances in which a threaded fastener alone is useless. The shield or expansion anchor inserted in a predrilled hole expands when the bolt is driven into it and becomes wedged firmly in the hole, providing a secure base for the grip of the fastener.

CORRUGATED FASTENERS

The corrugated fastener is one way to fasten joints and splices in small boards. It is used particularly in the miter joint. Corrugated fasteners are made of sheet metal of 18 to 22 gauge with alternate ridges and grooves. The ridges vary from 3/16 to 5/16 inch center to center. One end is cut square, while the other is sharpened with beveled edges.

There are two types of corrugated fasteners (Fig. 4-21). One has the ridges running parallel, while the other has ridges running at a slight angle

to one another. The latter type has a tendency to compress the material since the ridges and grooves are closer at the top than at the bottom. These fasteners are made in several different lengths and widths. The width varies from 5/8 to 1 1/8 inches, and the length varies from 1/4 to 3/4 inch.

The fasteners are also made with different numbers of ridges, ranging from three to six ridges per fastener. Corrugated fasteners are used in a number of ways: to fasten parallel boards together, to make any type of joint, and as a substitute for nails where nails may split the timber. The fasteners have a greater holding power than nails in small timber and are often used in the installation of solid lumber paneling. The proper method of using corrugated fasteners is shown in Fig. 4-22.

ADHESIVES

There are many ways to bond solid lumber panel-

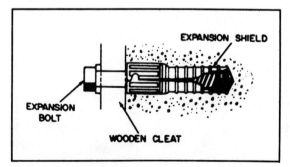

EXPANSION SHIELD

EXPANSION BOLT

WOODEN CLEAT

Fig. 4-20. Expansion bolt.

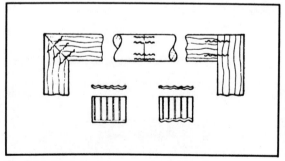

Fig. 4-22. Typical uses of corrugated fasteners.

ing to another surface. One of the oldest materials used for fastening is *glue*. Museums display furniture that was assembled with glue hundreds of years ago and is still in good condition. Good glue applied properly will form a joint which is stronger than the wood itself.

There are several classes of glue. Probably the best one for joint work and solid lumber paneling installation is animal glue. It may be obtained commercially in a variety of forms—liquid, ground, chipped, flaked, powered, or formed into sticks. The best grades of animal glue are made from hides. Some of the best bone glues, however, may give results as good as the low grades of hide glue.

Fish glue is a good all-around woodshop glue, but it is not as strong as animal glue. It is usually made in liquid form, and it has a disagreeable odor.

Vegetable glue is manufactured by a secret process for use in some veneering work. It is not a satisfactory glue for wood joints.

Casein glue is made from milk in powdered form. The best grades of casein glue are water-resistant and are, therefore, excellent for forming waterproof joints. Casein glue, however, doesn't adhere well to oak. To join oak surfaces with it, coat the wood with a 10-percent solution of caustic soda and allow it to dry. Then apply the casein glue to form a strong joint.

Blood albumin glue is also practically waterproof, but to use it you need very expensive equipment. It is, therefore, not often used for panel installation.

Plastic resin glue comes in liquid or powder form. It is durable and water resistant, but like casein glue, it doesn't adhere too well to oak. Plastic resin glue is used in the manufacture of balsa wood and plywood components.

Each type of glue must be prepared and used in a special manner if you are to get the strongest possible joint. Instructions are always found on the label of the container. Study them carefully before you attempt to use the glue.

A lot depends on the wood itself. Dry wood makes stronger joints than wood which is not well seasoned. This is easy to understand if you'll remember that water in the wood will decrease the amount of glue which can be absorbed.

Contact Adhesives

Without a doubt, the most difficult method of adhesive installation for the average homeowner is the use of contact cement. Unless you are well versed in the use of contacts, this method of application is best left up to the professionals.

Various manufacturers have different instructions, normally set forth on their labels, which should be closely followed. The rubber-base contact adhesives are applied to both surfaces which will be joined together when panels are in place.

There are two very important points in contact cement application, either one of which can create a considerable amount of difficulty. First, the adhesive applied to panels must be permitted to set a certain length of time between coats and prior to application. Either too much or too little time will create a faulty contact of panel to surface. Secondly, there is no room for error in adjusting panels to fit the wall area once the adhesive has been applied. It is impossible to move the panels to adjust to floor, or ceiling, or electrical or heat outlets once the contact has been made. The contact cement method of application, however, has the advantage of eliminating face nailings.

Panel Adhesives

A satisfactory application of solid lumber paneling can be obtained with a minimal amount of nailing through the use of panel adhesives. Where furring strips have been placed against the wall area to be covered, the adhesive should be applied immediately before the application of the panels. By nailing the panel to the furring strips with a brad or nail, which is then countersunk, sufficient support is given to the board to permit the adhesive to cure (approximately 24 hours). In this manner both an adhesive and nail application is made, and an extremely secure application is obtained.

Installation of solid lumber paneling without the

use of any nails whatever may be done with panel adhesives in two ways:

☐ Mechanical blocking methods can hold the panel strips in place for a sufficient length of time to permit the adhesive to cure (usually overnight). For example, braces can be devised using temporary support from the opposite wall of a room.

☐ A 2 × 4 can be nailed into the wall at top and bottom only, where molding will conceal holes. Be careful that there is no middle board warp.

Several manufacturers are now supplying adhesives which are laid on in thick beads by a caulking gun (Fig. 4-23) in order to apply solid lumber panel strips directly to the wall surfaces or furring strips. Uneven depressed areas in the wall should be marked to receive a heavier amount of adhesive to help bridge the gap between panel and surface. The panels are applied to the beads of mastic and pressed lightly, making certain the panels are plumb. Contact is made and the panels are now ready for molding.

This one-step operation levels the face of irregular and subsurfaces as a result of the buildup of bead and adheres the panels. Always follow the manufacturer's application procedure when using contact cement or panel adhesives.

RULES FOR GLUEING

Keep these rules in mind as you select and plan solid lumber paneling installation with adhesives (Fig. 4-24).

☐ Don't sandpaper surfaces to be glued. A rough surface with pores allows the glue to adhere better.

☐ Before applying the glue, make certain that the wood surfaces to be glued fit tightly together. Any old glues, chemicals, paint, varnish, or other loose elements should be removed.

☐ If you are using a brush to install the adhesive, make sure the glue is thin enough to run freely from the brush. Thin if necessary.

☐ Be sure that the area in which you will be applying the adhesive is at room temperature—55 to 75 degrees Fahrenheit, or as recommended by the manufacturer.

☐ Apply pressure to the panels once they are

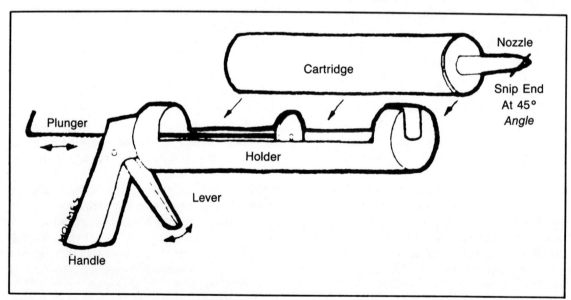

Fig. 4-23. Caulking gun used for applying adhesive.

Fig. 4-24. Using adhesives to install paneling (courtesy Georgia-Pacific).

in place so they will set. Make any last minute adjustments for a tight fit. Some types of glue will require continued pressure for 2 to 24 hours.

You've planned your solid lumber paneling project, selected the paneling, prepared it for installation, and chosen the fastener or adhesive you will use to install the panels. Everything comes together in Chapter 5 as you learn how to install solid lumber paneling.

Chapter 5

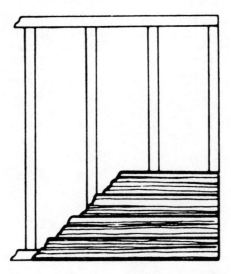

Installing Solid Lumber Paneling

VARIOUS TYPES AND PATTERNS OF WOODS can be applied to walls to obtain desired decorative effects. For informal treatment, knotty pine, white-pocket Douglas fir, sound wormy chestnut, and pecky cypress—finished natural or stained and varnished—may be used to cover one or more sides of a room.

Solid lumber paneling should be thoroughly seasoned to a moisture content near the average it reaches in service—in most areas about 8 percent. Allow the material to reach this condition by placing it around the wall of the heated room.

Boards may be applied horizontally or vertically, but the same general methods of application should pertain to each. The following is a guide to applying matched solid lumber paneling. Once you've learned the basic steps to installation, we'll cover the variations and details.

☐ Apply solid lumber paneling over a vapor barrier and insulation when application is on the exterior wall framing or blocking (Fig. 5-1).

☐ Boards should not be wider than 8 inches, except when a long tongue or matched edges are used.

☐ Thickness should be at least 3/8 inch for 16-inch spacing of frame members, 1/2 inch for 20-inch spacing, and 5/8 inch for 24-inch spacing.

☐ Maximum spacing of supports for nailing should be 24 inches on center (blocking for vertical applications).

☐ Nails should be five penny or six penny (5d or 6d) casing or finishing nails.

Use two nails for boards up to 6 inches wide and three nails for boards 8 inches or wider. One nail can be blind-nailed in matched paneling.

Wood paneling in the form of small wood squares can also be used for an interior wall covering. When used over framing and a vapor barrier, blocking should be so located that each edge has full bearing. Each edge should be fastened with casing or finishing nails. When two sides are tongued and grooved, one edge (tongued side) may be blind-

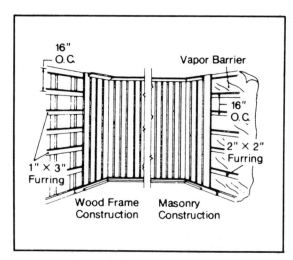

Fig. 5-1. Applying solid lumber paneling over a vapor barrier (courtesy Champion International).

nailed. When wide solid lumber paneling (16 by 48 inches or larger) crosses studs, it should also be nailed at each intermediate bearing. Matched (tongued-and-grooved) sides should be used when no horizontal blocking is provided or paneling is not used over a solid backing.

These are the basic considerations. Let's look closer at the elements of solid lumber panels.

PRECONDITIONING

To minimize the amount of expansion and contraction which could occur from changes in moisture content, remove individual pieces of lumber paneling from any package or wrapping and allow them to acclimate to room conditions for at least 48 hours before they are installed. Stack the panels with lumber or spacers between layers and a space between individual pieces (Fig. 5-2). If high humidity exists, allow a longer period for preconditioning.

Preconditioning is a very important element to the installation of solid lumber paneling because it allows the wood time to adjust to its new surroundings. Wood that has sat outside during the wet season may have a high moisture content. If it is brought into a dry room and immediately installed, it will begin to "sweat" and shrink, causing numerous problems for the wood and any finish. Stacking for preconditioning also allows the location and opportunity for treating or finishing the solid lumber paneling should you decide to do so prior to installation (see Chapter 6).

PREPARING WALLS

It's time to begin. First paint, if desired, the ceiling, baseboards, molding, trim, and window and door casings to remain with your new paneling. Remove those to be eliminated or replaced with the solid lumber paneling. If your walls are in good condition, you can glue and nail solid lumber paneling directly into studs, butting neatly against existing trim (Figs. 5-3 through 5-5).

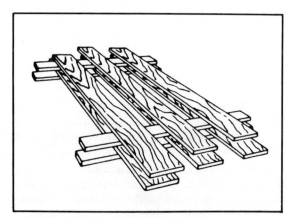

Fig. 5-2. Stacking solid lumber paneling with spacers (courtesy Champion International).

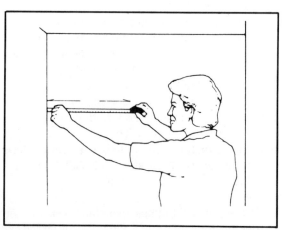

Fig. 5-3. Measuring walls to find studs (courtesy Georgia-Pacific).

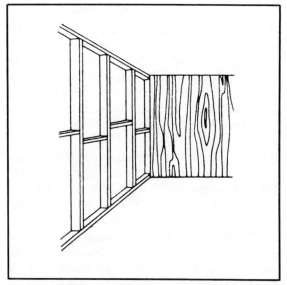

Fig. 5-4. Installation of decorative paneling over wall studs (courtesy Hardwood Plywood Manufacturers Association).

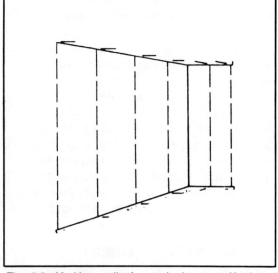

Fig. 5-5. Marking walls for studs (courtesy Hardwood Plywood Manufacturers Association).

On uneven, badly cracked, or very rough walls, solid lumber can be nailed to a simple framework of 1-×-4 or 1-×-3 furring strips of any wood species and grade that are kiln dried (Fig. 5-6). Fitting shims (ordinary wood shingles) under the strips will even up severe surface gaps.

To figure furring strip requirements for vertical paneling, measure wall width to determine the length of one horizontal strip and wall height to figure how many strips you'll need, spaced about 24 inches on center. Furring strips must also be nailed around windows, doors, and all other openings; so add these linear measurements to the total. At windows and doors, nail strips flush with inside or wall edges of jambs so that built-up strips can be added later.

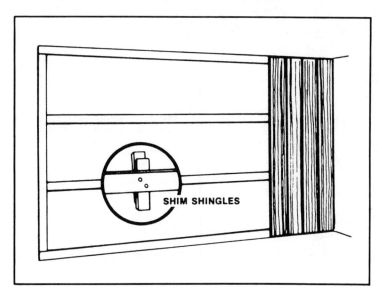

SHIM SHINGLES

Fig. 5-6. Preparing walls with shim shingles under furring (courtesy California Redwood Association).

95

ELECTRICAL OUTLETS

Next, remove all outlet and switch plates (tape screws to each to avoid losing them) and hot and cold air duct covers. If you want to extend outlet boxes for added paneling thickness, shut off electricity at the fuse panel beforehand.

Cut openings for heater ducts and electrical outlets as individual boards go up (Fig. 5-7). To mark the openings, apply lipstick to the edges of the electrical box or heater duct, align the panel, and press into place. Removal of the panel will leave an exact print where the hole should be cut. Drill four large holes just inside the marked corners and saw from the corners with a saber or keyhole saw.

ARRANGE LUMBER

The third step in solid lumber paneling installation is to arrange the lumber. Before you nail or glue paneling, take time to arrange boards carefully against the wall according to desired grain and color patterns. For example, the most attractive boards can be placed so furniture won't hide them, and sapwood-streaked lumber can be grouped to exaggerate flare effects. This is your chance to be creative. Refer to Fig. 5-8.

VERTICAL PANELING

Begin vertical paneling at an inside corner and work left to right if you're right-handed, and right to left if you're left-handed. Keep groove edges toward the starting corner and tongue edges toward your work direction. Trial-fit the first board, check for plumb (perpendicular to the floor), then nail with 5d or 6d finishing nails, even if other boards are to be glued on.

On furring strips, face-nail top and bottom, and blind-nail (Fig. 5-9) through the tongue at each strip. Without furring strips, face-nail top, bottom, and every 2 feet about 3/4 inch from the corner edge, angling the nail to penetrate studs.

To avoid splitting board tongues and ends when nailing, you can prebore nail holes with a drill bit about 3/4 the diameter of the nail shank. To avoid hammer marks, drive nails within 1/8 inch of the board surface and finish with a nail set.

Nail holes can be left as is or filled in with a colored wood putty or filler. A mixture of wood sander dust and clear cement forms a quick-drying filler. Oil-based fillers stain the adjoining wood and should be used only if you plan to apply an oil-based finish to the paneling.

Measure all other boards carefully and trial fit.

Fig. 5-7. Making cutouts in solid lumber paneling for electrical outlets (courtesy California Redwood Association).

Fig 5-8. Arrange lumber for the greatest use of grain patterns and coloration (courtesy California Redwood Association).

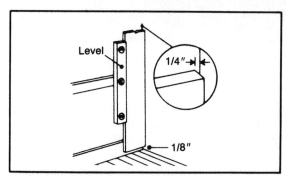

Fig. 5-9. Aligning the first panel strip (courtesy Champion International).

Tap them into place with a hammer and tapping block—a scrap with the groove edge intact to fit over board tongues (Fig. 5-10). Check for plumb and, if necessary, slightly angle groove-to-groove fitting to make it square. When butt-joining short board ends together in the middle of a wall, be sure the joint falls over a stud, blocking, or furring strip.

At an outside corner, miter-join lumber with a table saw, making 45-degree cuts on board edges. Without a table saw, you can butt-join, trim off groove or tongue edges squarely, sand finely, and cover the joint with trim or molding, if desired.

Fig. 5-10. Tap solid lumber panel strips into place with a scrap block (courtesy California Redwood Association).

With a hand power saw, trim lumber face down; with a conventional hand saw, trim face up to prevent face chip-out. On furring strips, face-nail paneling at top, bottom, and down the middle. Without furring, face-nail at top, bottom and every 2 feet, about 3/4 to 1 inch from the corner edge.

If you glue panels, apply adhesive generously with a caulking gun to the back of a prefitted board (Fig. 5-11). Let it set according to package directions, then hold the board to the wall so both are coated. Remove it (Fig. 5-12), wait again, then tap the board into place and face-nail top and bottom with 8d finishing nails (two nails at each end for wider boards). Because of waiting periods, you can finish one board while adhesive sets on another.

The last board may have to be trimmed to fit

into a corner. Angle trim the board's corner edge slightly with a block plane, with the wide part of the angle toward the wall (Fig. 5-13).

ADDING MOLDINGS AND BASEBOARDS

To install moldings and baseboards, measure floors and ceilings separately. Miter-join molding and baseboard ends on outside corners (Fig. 5-14) and butt-join them on inside corners (Fig. 5-15). A patterned molding must be miter-joined on inside corners also.

WINDOW AND DOOR TRIM

Before applying furring strips or paneling you

Fig. 5-11. Apply glue with caulking gun (courtesy California Redwood Association).

Fig. 5-12. Make sure adhesive is on both wall and panel (courtesy California Redwood Association).

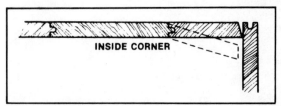

Fig. 5-13. Installing the last board in the corner (courtesy California Redwood Association).

removed window and door trim or casings. After paneling, finish doors and windows, nailing build-up strips into jambs to cover board ends and/or furring strip edges where necessary.

Start vertical and diagonal paneling from an inside corner and work over to a window or door (Figs. 5-16 through 5-18). To fit diagonal paneling at opening edges, miter-cut board ends. At corners

Fig. 5-14. Miter joint molding (courtesy California Redwood Association).

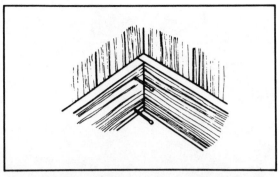

Fig. 5-15. Butt joint molding at inside corner (courtesy California Redwood Association).

Fig. 5-16. Start installing paneling around window by arranging and matching panels (courtesy California Redwood Association).

Fig. 5-17. Apply adhesive to the back of the panels (courtesy California Redwood Association).

of openings, position a board diagonally as it will be in paneling, mark horizontal and vertical edges of the corner, make a cut, and finish with a wood rasp for a close fit. Start horizontal paneling from the baseboard or floor and work up.

Figure 5-19 shows how to install door trim without furring strips. Fit board edges flush with the jambs; you don't need build-up strips. When fitting board ends around openings, leave space to nail a build-up strip into the jamb flush with paneling and the jamb.

Figure 5-20 illustrates the installation of door trim with furring strips. Fit board ends and edges flush with furring strips. Nail into the jamb a build-up strip equal to the thickness of the furring plus paneling.

Fig. 5-18. Tacking on molding after the paneling is in place (courtesy California Redwood Association).

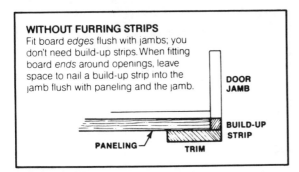

WITHOUT FURRING STRIPS
Fit board *edges* flush with jambs; you
don't need build-up strips. When fitting
board *ends* around openings, leave
space to nail a build-up strip into the
jamb flush with paneling and the jamb.

DOOR JAMB

BUILD-UP STRIP

PANELING

TRIM

Fig. 5-19. Installing door trim without furring strips (courtesy California Redwood Association).

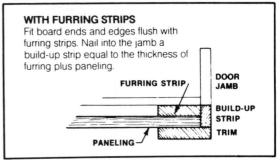

WITH FURRING STRIPS
Fit board ends and edges flush with
furring strips. Nail into the jamb a
build-up strip equal to the thickness of
furring plus paneling.

FURRING STRIP

DOOR JAMB

BUILD-UP STRIP

TRIM

PANELING

Fig. 5-20. Installing door trim with furring strips (courtesy California Redwood Association).

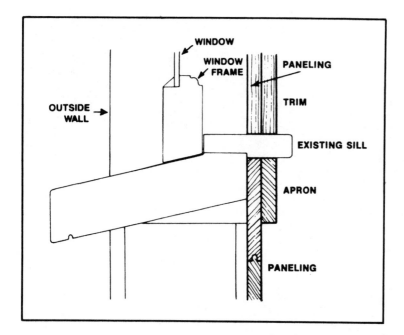

Fig. 5-21. Installing solid lumber paneling window trim (courtesy California Redwood Association).

Figure 5-21 offers the details for installing solid lumber paneling as window trim around an existing sill. Figure 5-22 shows how a false sill can be added to match the paneling.

HORIZONTAL PANELING

Though vertical paneling is very popular, many people prefer to install solid lumber paneling in a horizontal pattern. Let's take a look at how this can

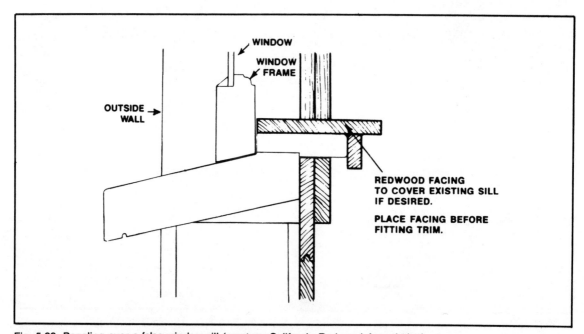

Fig. 5-22. Paneling over a false window sill (courtesy California Redwood Association).

Fig. 5-23. Installing horizontal solid lumber paneling (courtesy California Redwood Association).

be easily done by the do-it-yourselfer.

To even up wall surfaces for horizontal paneling, apply vertical furring strips. Measure wall height to find the length of one strip and wall width to determine the number of strips needed, spaced on wall studs or about 24 inches apart. Nail furring strips around doors, windows, and all other openings.

Nail the baseboard first (a 1 × 6 or 1 × 8) with 6d finishing nails at each stud or furring strip. Then butt the groove edge of the first board against the baseboard, face-nail the ends, and blind-nail at studs. With wider lumber, face-nail about 3/4 inch

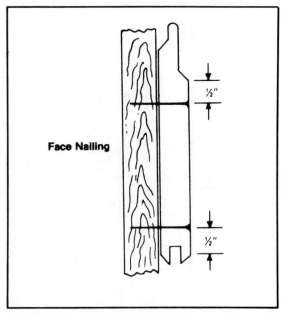

Face Nailing

Fig. 5-25. Face-nailing solid lumber paneling (courtesy Champion International).

from the baseboard, one nail per stud. Without a baseboard, place the first board tight against the floor and nail as described (Fig. 5-23). Figures 5-24 through 5-28 show fastening methods.

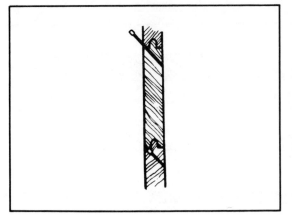

Fig. 5-24. Nailing paneling (courtesy California Redwood Association).

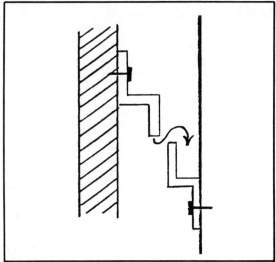

Fig. 5-26. Two-inch clips can be used to hang paneling (courtesy Hardwood Plywood Manufacturers Association).

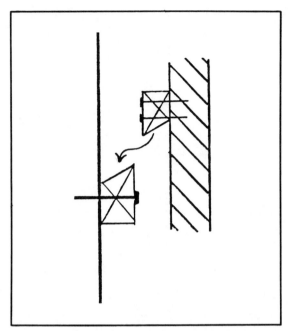

Fig. 5-27. Using tapered wooden blocks to hang solid lumber paneling (courtesy Hardwood Plywood Manufacturers Association).

On an outside corner, miter the ends, making a cut greater than 45 degrees in order to fit tightly (Fig. 5-29); or butt-join, covering with molding or trim (Fig. 5-30). Butt-join inside corners also. Handle baseboards the same way, but without molding or trim. To fit the last board at the ceiling, trim the corner edge slightly and at an angle with a block

plane, with the wide part of the angle toward the wall. Figure 5-31 illustrates three suggested butt joint corner trim methods for horizontal paneling.

DIAGONAL PANELING

Diagonal paneling (Fig. 5-32) is often used for accent walls rather than whole rooms. You'll need 15 percent more lumber. For instance, paneling 100 square feet of wall space with 6-inch wide lumber requires 132 1/4 square feet (115 × 1.15 = 132 1/4) or 264 1/2 linear feet (132 1/4 × 2 = 264 1/2), plus 5 percent for errors.

If walls need them, apply furring strips vertically, as for horizontal paneling. Start paneling at an inside corner, working left to right if you're right-handed, and right to left if you're left-handed, as for vertical paneling. Keep the groove edge toward the starting corner and the tongue edge toward the work direction.

Arrange your first three boards flat on the floor so that on the wall tongue edges will be up. Fit tongues in grooves tightly, position a carpenter's square, and mark for a 45-degree miter cut (Fig. 5-33) across all three boards. Saw and trial fit. Gently shaving the ends will ensure a tighter fit. Then face-nail the corner piece and blind-nail the rest at each end and at the intersecting stud or furring strip (Fig. 5-34).

To butt-join board ends in the middle of a wall you can trim square or miter-cut. The joint should

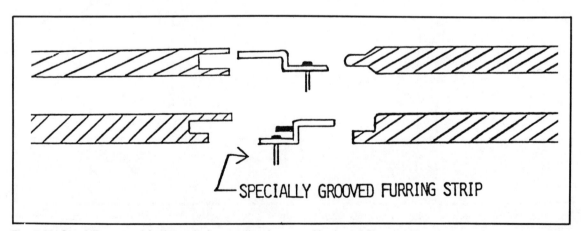

SPECIALLY GROOVED FURRING STRIP

Fig. 5-28. Specially grooved furring strip for paneling (courtesy Hardwood Plywood Manufacturers Association).

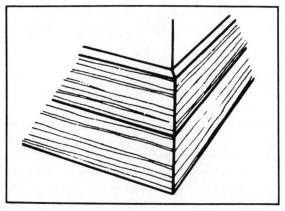

Fig. 5-29. Miter joint for outside trim (courtesy California Redwood Association).

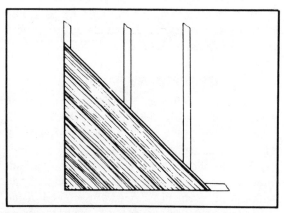

Fig. 5-32. Installing diagonal solid lumber paneling (courtesy California Redwood Association).

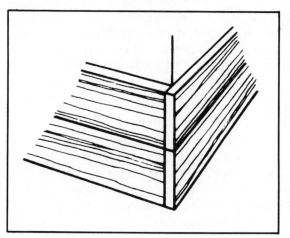

Fig. 5-30. Butt joint for outside trim (courtesy California Redwood Association).

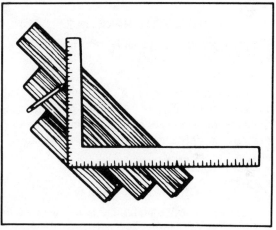

Fig. 5-33. Diagonal paneling starts with a good square corner (courtesy California Redwood Association).

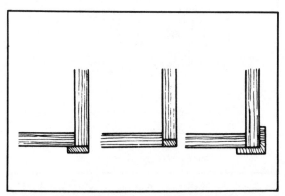

Fig. 5-31. Three butt joint corner trim methods for horizontal paneling (courtesy California Redwood Association).

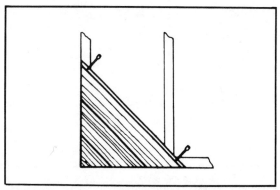

Fig. 5-34. Face-nail the corner piece and blind-nail the rest (courtesy California Redwood Association).

107

cover a stud, blocking, or furring strip. Butt-join paneling at corners, covering outside edges with trim or molding, if desired.

HERRINGBONE PANELING

One popular solid lumber paneling pattern is called the herringbone (Fig. 5-35) because that's what it looks like. It's applied much like the diagonal paneling just discussed.

Plan to start from an inverted V location mid-wall. Lay out the herringbone pattern so that odd widths will be equally divided at each end of the wall for a balanced appearance. Leave the same spaces recommended for straight diagonal applications.

WAINSCOT PANELING

Wainscot paneling (Fig. 5-36) is returning to vogue, especially with older style homes. Installation is easy.

Install lumber furring horizontally at the wainscot height desired. All other recommendations for wall applications over furring apply. Cap the top edge of the wainscot with a wood chair rail or end cap molding.

CEILING PANELING

You can also install solid lumber paneling on other interior surfaces, such as the ceiling of a room. In fact, ceiling paneling is becoming increasingly popular.

Face-nail paneling directly to ceiling joists over gypsum board into joists or lumber furring spaced not more than 16 inches on center. For diagonal and herringbone applications, face-nail paneling over lumber furring spaced not more than 12 inches on center. Don't use adhesive for ceiling applications. Leave a minimum gap of 1/4 inch around the perimeter of the ceiling and conceal with moldings. Other than this, installing solid lumber ceiling paneling is almost identical to installing wall paneling.

SPECIAL INSTALLATIONS

It would be so simple to install solid lumber paneling on four flat walls and be done. It wouldn't be as decorative, however, as installing it in alcoves, arches, coves, stairwells, and other places. So let's look at how to solve these special installation problems.

Alcoves

Look at an alcove as being a miniature three-sided room. Then you can go ahead and treat it with solid lumber paneling and moldings in much the same way. Don't overlook other ways to use the space, however. Consider building in a window seat or desk, or perhaps a day bed or full-size bed. An

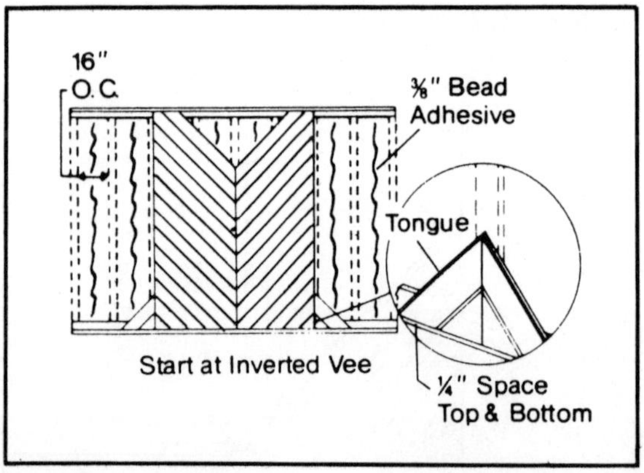

Fig. 5-35. Herringbone solid lumber paneling (courtesy Champion International).

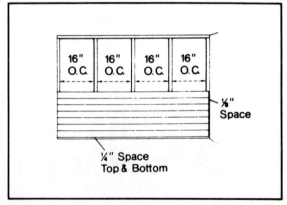

Fig. 5-36. Wainscot paneling (courtesy Champion International).

alcove doesn't have to be a walk-in space to be functional and attractive.

Paneling outside the arch is simply a matter of measuring and cutting the paneling to fit the area above the curve. Paneling the underside of the arch—and applying the necessary molding—can be trickier. If the adjoining room is painted or papered and will remain so, your best answer may be to make your transition right then and there, avoiding the underside of the arch.

Where paneling is to be carried through both rooms, you will need to bend a strip of plywood paneling of a matching texture and design to the underside curve. *Caution*: use only plywood-based paneling (Fig. 5-37), since wood fiber substrate paneling may break if you try to curve it to the arch.

Fig. 5-37. Installing plywood-base paneling (courtesy Georgia-Pacific).

Experiment with a scrap strip of paneling to test its flexibility.

Measure how much paneling is needed. The strip's width will generally range from 4 to 8 inches. You can choose to make the end joints meet at the top of the arch or at the wainscot level. Select a strip without any grooves for improved break-resistance. If the paneling is not flexible enough to conform to the arch without help, carefully make a series of saw kerfs across the back of the strip. Use a 1/8-inch blade and make your cuts no more closely spaced than needed to achieve a smooth curve.

When the strips are cut and ready to install, apply them to the arch with adhesive following the manufacturer's instructions. Use enough nails to hold the strips in place while the adhesive dries.

Standard 3/4-inch outside-corner molding is usually flexible enough to conform to the curve, but may look skimpy and out of proportion if used to frame a large opening. A better solution is to cut curved sections 2 1/2 to 3 inches wide from a 1- × -6 pine board, then stain or paint.

There's a simpler, and sometimes more dramatic solution to the whole curved-arch riddle that may appeal to you. Square off the corners with 45-degree straight sections and flatten the top, using 1- × -6 trim boards to frame and line the opening.

Curved Plaster Cove Ceilings

Let's consider installation around curved plaster cove ceilings. Since curved-cove ceilings are usually in rooms with walls higher than 8 feet, the quickest solution is to panel to the 8-foot level and finish off with cap molding. Then paint the wall above to match and merge with the ceiling.

A more interesting and creative solution is to install cove lighting. Box in the cove area at the ceiling line with a simple paneled-lumber framework. Now you have the option of installing indirect perimeter lighting (bouncing light off the ceiling) or wiring the box for built-in spotlights to "wash down" the walls, accenting artwork and furnishings.

Attics

An unfinished attic can yield several extra rooms with the help of dimensional lumber and solid lumber paneling (Fig. 5-38 through 5-40). Short side walls may already be in place, or you may have to build these walls with a stud framework tied to the roof rafters and floor. To avoid bumping your head, a minimum wall height of 5 1/2 to 6 feet is recommended. Architects often refer to these areas as *kneewalls*. You can use the space behind the kneewalls to build in bookcases, cupboards, or storage drawers. You can also use the space close to the walls in ways that don't require much headroom—a built-in desk, cushioned benches, knick-knack or TV shelves, and so on. If you're fortunate to have a dormered window, make a minialcove with built-in window seat. Easy and fancy solid lumber paneling projects will be covered later in this book.

You may find that existing walls have studs on 16-inch centers, while roof rafters are on 24-inch centers. Check it out before you begin. Begin by

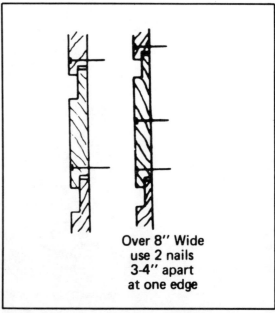

**Over 8″ Wide
use 2 nails
3-4″ apart
at one edge**

Fig. 5-38. Installing tongue-and-groove solid lumber paneling.

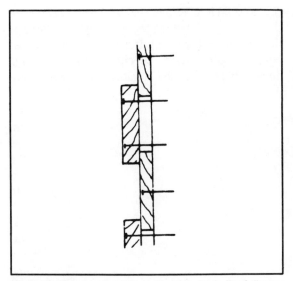

Fig. 5-39. Board-on-board paneling can be built of dimensional lumber.

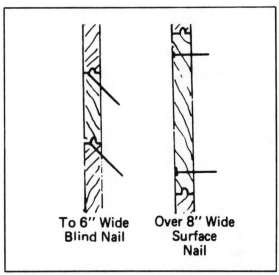

To 6" Wide
Blind Nail

Over 8" Wide
Surface
Nail

Fig. 5-40. Blind-nailing and surface-nailing tongue-and-groove paneling.

paneling the ceiling. Place the bottom edge of the panels where the roof meets the side wall. Now panel the walls. Moldings may be used to cover the joints where ceiling and walls meet. *Note*: before paneling, don't miss the opportunity to install sufficient insulation in walls and ceilings to ensure comfort and conserve energy.

Stairwells

Cutting solid lumber paneling to fit accurately around a basement stair wall is not difficult. In many cases, vertical stair risers are flush to the stringer side (to which your panels will be attached), but the horizontal stair treads overhang by several inches. You must, therefore, fit around the treads. One way of doing so is to make a paper pattern from large pieces of paper. Just tape the paper over the stair area and trim it into an accurate pattern (under the projecting stair treads). Carefully remove your pattern from the stair and transfer the outline to a row of solid lumber panels assembled, but not nailed, on the floor or workbench. Then carefully cut the panels to size with a saber saw. Test them against the stair for accuracy, then apply with nails or adhesive as outlined earlier in this chapter.

Stone Walls and Fireplaces

When paneling will butt up against a stone or brick fireplace or wall, you'll need to do an accurate fitting job. In this case, the compass method is the best method. Tack one panel strip temporarily in place and scribe a line parallel to the fireplace edge with a pencil. Cut the panel along the scribe line with a saber or similar saw.

Paneling Over Masonry Walls

Special problems arise when you decide to install solid lumber paneling over masonry walls. Rather than attempt to use anchor nails throughout, many do-it-yourselfers simply install a furring wall between the masonry and the panel strips.

Actual furring walls can easily be installed, as outlined earlier, with 1- x -2 lumber, or 3/8- to 1/2-inch plywood strips cut 1 1/2 inches wide. Again, shims can be used to true-up, or level, the wall first. Make sure that the installation pattern of your furring wall doesn't contradict the pattern of your solid lumber paneling wall.

If necessary, you can drill a hole with a carbide-tipped bit, insert wooden plugs or expansion shields, and nail or bolt the shimmed frame in place.

Then you'll have a sturdy and useful surface on which to install your solid lumber paneling.

FINISHING UP

As you stand back to appreciate your newly installed solid lumber panelled wall, you may decide that it needs a finishing touch. That's the topic of the Chapter 6 Finishing. It's also the last chapter before you're given specific instructions for numerous practical solid lumber paneling projects.

Chapter 6

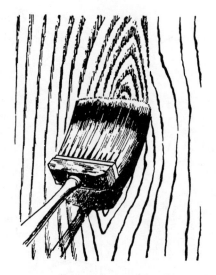

Finishing

O NCE YOU'VE INSTALLED YOUR SOLID LUM-
ber paneling, 80 percent of the job is done.
The other 20 percent, however, the trim and finish
can increase the beauty of your project many times.
In this chapter, we'll take a closer look at how to
select and install various trims and finishes to solid
lumber paneling so that your project is exactly what
you want it to be.

The decorative treatment for interior walls,
doors, trim, and other millwork may be paint or a
natural finish with stain, varnish, or other non-
pigmented material. The type of paint or natural
finish desired often determines the type and species
of wood to be used. Interior finish that is to be
painted should be smooth, close-grained, and free
from pitch steaks. Some species having these requirements in a high degree include ponderosa pine,
northern white pine, redwood, and spruce. When
hardness and resistance to hard usage are additional
requirements, species such as birch, gum, and
yellow poplar are good choices for decoration and
trim.

For natural finish treatment, a pleasing figure,
hardness, and uniform color are usually desirable.
Species which meet these requirements include ash,
birch, cherry, maple, oak, and walnut. Some require
staining for best appearance.

DOOR TRIM

Figures 6-1 and 6-2 show home components, and
Fig. 6-3 shows the installation of trim over a typical
doorframe. The trim or casings are nailed to both
the jamb and the framing studs or headers, allowing about a 3/16-inch edge distance from the face
of the jamb. Finish or casing nails in 6d or 7d sizes,
depending on the thickness of the casing, are used
to nail into the stud. Finishing nails in 4d or 5d sizes
or 1 1/2-inch brads are used to fasten the thinner
edge of the casing to the jamb. In hardwood, it is
usually advisable to predrill to prevent splitting.
Nails in the casing are located in pairs and spaced
about 16 inches apart along the full height of the
opening and at the head jamb.

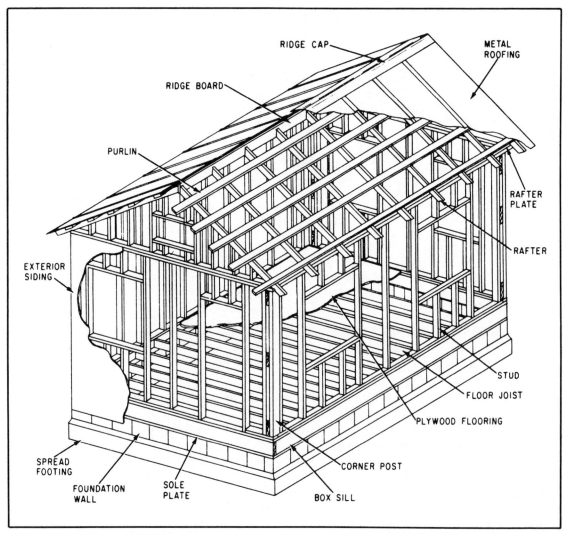

Fig. 6-1. Typical light-frame construction.

Casings with any type of molded shape must have a mitered joint at the corners (Fig. 6-4). When the casing is square-edged, a butt joint may be made at the junction of the side and head casing (Fig. 6-5). If the moisture content of the casing is well above the recommended percentage, a mitered joint may open slightly at the outer edge as the material dries. This opening can be minimized by using a small glued spline at the corner of the mitered joint. Actually, use of a spline joint under any moisture condition is considered good practice, and some prefitted jamb, door, and casing units are provided with splined joints. Nailing into the joint after drilling will aid in retaining a close fit.

WINDOW TRIM

The casing around interior window frames should be the same pattern as that used around the interior doorframes. Other trim used for a double-hung window frame includes the sash stops, stool, and apron (Fig. 6-6). Another way of using trim around windows is to enclose the entire opening with casing (Fig. 6-7). The stool is then a filler member between the bottom sash rail and the bottom casing.

The *stool* is the horizontal trim member that laps the window sill and extends beyond the casing at the sides, with each end notched against the wall. The *apron* serves as a finish member below the stool.

The windowsill is the first piece of window trim to be installed. It is notched and fitted against the edge of the jamb and the wall line, with the outside edge being flush against the bottom rail of the window sash. The stool is blind-nailed at the ends so that the casing and the stop will cover the nailheads. Predrilling is usually necessary to prevent splitting. The stool should also be nailed at

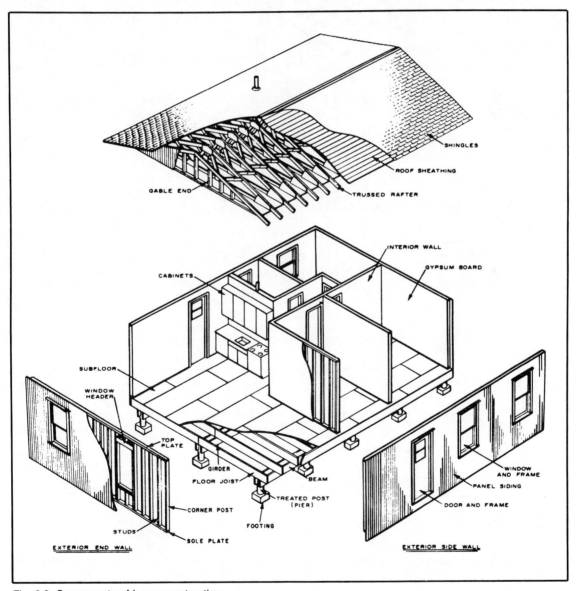

Fig. 6-2. Components of home construction.

115

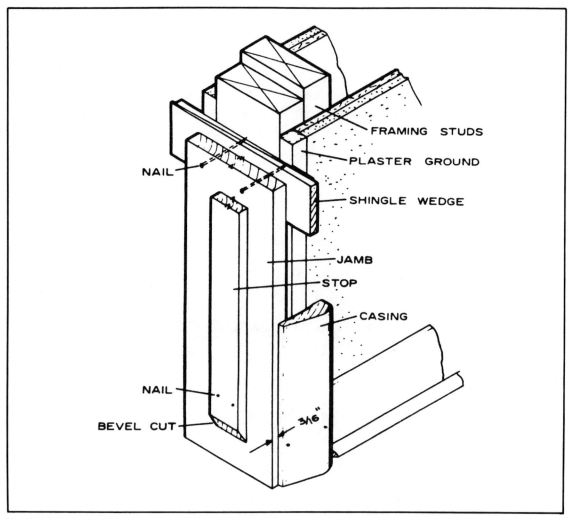

Fig. 6-3. Installation of trim over a typical doorframe.

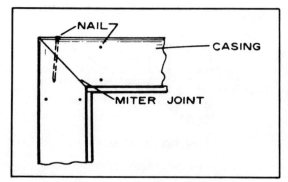

Fig. 6-4. Corner casings with mitered joints.

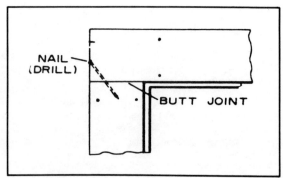

Fig. 6-5. Butt jointing.

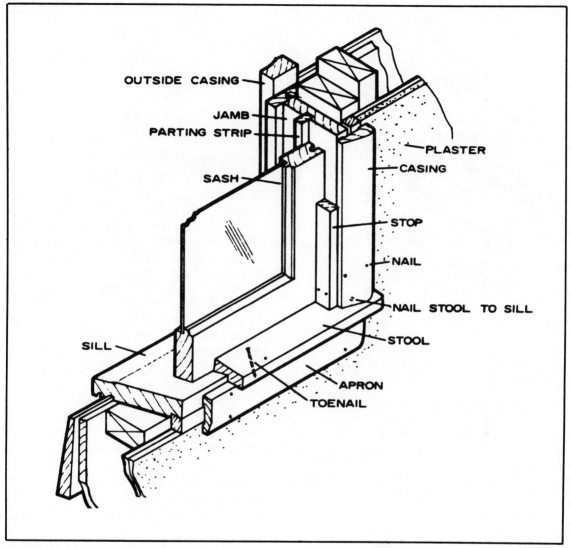

OUTSIDE CASING

JAMB

PARTING STRIP

SASH

PLASTER

CASING

STOP

NAIL

NAIL STOOL TO SILL

STOOL

SILL

APRON

TOENAIL

Fig. 6-6. Trim for double-hung windows.

midpoint to the sill and to the apron with finishing nails. Face-nailing to the sill is sometimes substituted or supplemented with toenailing of the outer edge to the sill.

The casing is applied and nailed as described for doorframes, except that the inner edge is flush with the inner face of the jambs so that the stop will cover the joint between the jamb and the casing. The window stops are then nailed to the jambs so that the window sash slides smoothly. Channel-type weather stripping often includes full-width metal subjambs into which the upper and lower sash slide, replacing the parting strip. Stops are located against these subjambs instead of the sash to provide a small amount of pressure. The apron is cut to a length equal to the outer width of the casing line. It is nailed to the windowsill and to the 2- x -4 framing sill below.

When casing is used to finish the bottom of the window frame as well as the sides and top, the nar-

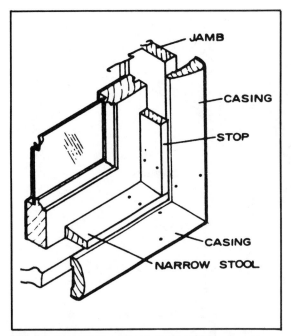

Fig. 6-7. Casing the entire window opening.

row stool butts against the side window jamb. Casing is then mitered at the bottom corners and nailed as previously described.

BASE MOLDINGS

Base moldings serve as a finish between the finished wall and floor. They are available in several widths and forms. A two-piece base consists of a

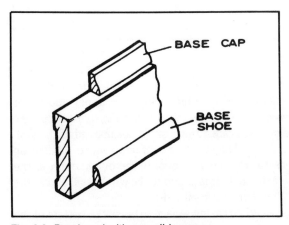

Fig. 6-8. Baseboard with a small base cap.

Fig. 6-9. One-piece base.

baseboard topped with a small base cap (Fig. 6-8). When your solid lumber panelled wall is not straight and true, the small base molding or base cap will conform more closely to the variations than will the wider base alone. A common size for this type of baseboard is 5/8 × 3 1/4 inches or wider.

A one-piece base varies in size from 7/16 × 2 1/4 inches (Fig. 6-9) to 1/2 × 3 1/4 inches (Fig. 6-10) and wider. Although a wood member is desirable at the junction of the wall and carpeting to serve as a protective "bumper," wood trim is sometimes eliminated entirely.

Most baseboards are finished with a base shoe, 1/2 × 3/4 inches in size (see previous figures). A one-base molding without the shoe is sometimes placed at the wallfloor junction, especially where carpeting might be used (Fig. 6-11 and 6-12).

Square-edged baseboard should be installed with a butt joint at the inside corners and a mitered joint at the outside corners (Fig. 6-13). It should be

Fig. 6-10. Large one-piece base.

Fig. 6-11. Preparing paneling for the molding (courtesy Georgia-Pacific).

Fig. 6-12. Carpeting is easier to install with a single-base molding (courtesy Georgia-Pacific).

nailed to each stud with two 8d finishing nails. A molded single-piece base, base moldings, and base shoe should have a coped joint at inside corners and a mitered joint at outside corners. A *coped joint* is one in which the first piece is square-cut against the wall or base and the second coped, which is accomplished by sawing a 45-degree miter cut and, with a coping saw, trimming the molding along the inner line of the miter (Fig. 6-14). The base shoe should be nailed into the subfloor with long slender nails, but not into the baseboard itself. Thus, if there is a small amount of shrinkage of the joists, no opening will occur under the shoe.

CEILING MOLDINGS

Ceiling moldings are sometimes used at the junction of wall and ceiling for an architectural effect

or to terminate solid lumber paneling or gypsum board (Fig. l6-15). As in the base moldings, inside corners should have a coped joint to ensure a tight joint and retain a good fit if there are minor moisture changes.

A cutback edge at the outside of the molding will partially conceal any unevenness of the plaster and make painting easier where there are color changes (Fig. 6-16). For narrow solid lumber paneling or for gypsum drywall, a small, simple molding might be more desirable (Fig. 6-17). Finish nails should be driven into the ceiling joists for large moldings when possible.

FINISHING WOOD

Wood and wood products in a variety of species, grain patterns, texture, and colors are available for

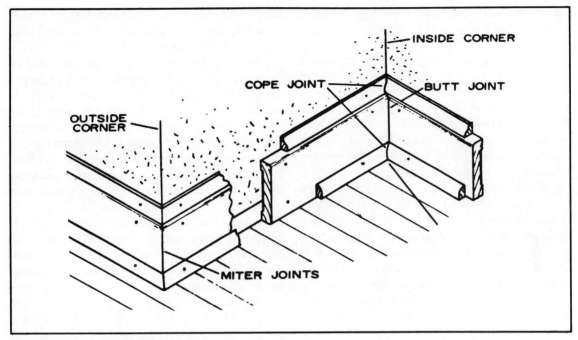

Fig. 6-13. Square-edged baseboard.

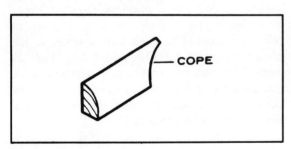

Fig. 6-14. Miter cuts.

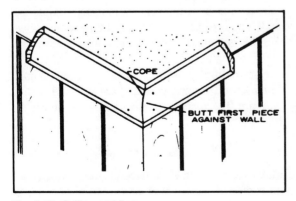

Fig. 6-15. Ceiling molding.

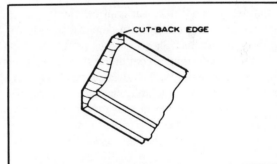

Fig. 6-16. Cutback edge.

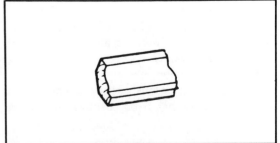

Fig. 6-17. Small simple molding.

use as solid lumber paneling. These wood surfaces can be finished quite effectively by several different methods.

Painting, which totally obscures the wood grain, is used to achieve a particular color decor. Penetrating preservatives and pigmented stains permit some or all of the wood grain and texture to show and provide a special color effect as well as a natural or rustic appearance. The type of finish, painted or natural, often depends on the wood to be finished and the desired result.

Wood surfaces shrink and swell the least are best for painting and finishing. For this reason, vertical- or edge-grained surfaces are far better than flat-grained surfaces of any species. Also, because the swelling of wood is directly proportional to density, low-density species are preferred over high-density species. Solid lumber in rooms such as the bathroom and kitchen are subjected to greater variations in moisture—and thus shrinking and swelling—than rooms with primary heat sources and little moisture input.

The properties of wood that detract from its paintability do not necessarily affect the finishing of such boards naturally with penetrating preservatives and stains. These finishes penetrate into wood without forming a continuous film on the surface. Therefore they will not blister, crack, or peel, even if excessive moisture penetrates the wood.

One way to further improve the performance of penetrating finishes is to leave the wood surface rough-sawn. Allowing the high-density, flat-grained wood surfaces or lumber to weather several months also roughens the surface and improves it for staining. Rough-textured surfaces absorb more of the preservative and stain, ensuring a more durable finish.

Interior finishing differs from exterior chiefly in that interior woodwork usually requires much less protection against moisture, more exacting standards of appearance, and a greater variety of effects. Good interior finishes used indoors should last much longer than paint coatings on exterior surfaces.

Opaque Finishes

Smoother surfaces, better color, and a more lasting sheen are usually preferred for interior woodwork, especially the wood trim. Enamels or semigloss enamels rather than paints are therefore used.

Before enameling, the wood surface should be extremely smooth. Imperfections such as planer marks, hammer marks, and raised grain, are accentuated by enamel finish. Raised grain is especially troublesome on flat-grained surfaces of the heavier softwoods. The hard bands of summerwood are sometimes crushed into the soft springwood in planing, and later are pushed up again when the wood changes in moisture content. It is helpful to sponge softwoods with water, allow them to dry thoroughly, and then sandpaper them lightly with sharp sandpaper before enameling. In new construction, woodwork should be allowed adequate time to come to its equilibrium moisture content before it is finished.

For hardwoods with large pores, such as oak and ash, the pores must be filled with wood filler before the priming coat is applied. The priming coat for all woods may be the same as for exterior woodwork, or special priming paints may be used. Knots in the white pines, ponderosa pine, or southern yellow pine should be shellacked or sealed with a special knot sealer after the primer coat is dry. A coat of knot sealer over white pines and ponderosa pine reduces pitch exudation and discoloration of light-colored enamels by colored matter present in the resin of the heartwood.

One or two coats of enamel undercoat are applied next. It should completely hide the wood and also present a surface that can easily be sanded smooth. For best results, the surface should be sanded before applying the finishing enamel; however this operation is sometimes omitted. After the finishing enamel has been applied, it may be left with its natural gloss or rubbed to a dull finish. When wood trim and paneling are finished with a flat paint, the surface preparation is not nearly as exacting.

Transparent Finishes

Transparent finishes are used on most hardwood and some softwood paneling and trim, according to personal preference. Most finishing consists of some combination of the fundamental operations of staining, filling, sealing, surface coating, or waxing. Before finishing, planer marks and other surface blemishes that would be accentuated by the finish may be removed.

Both softwoods and hardwoods are often finished without staining, especially if the wood has a pleasing and characteristic color. When it is used, however, stain often provides much more than color alone because it is absorbed unequally by different parts of the wood. It therefore accentuates the natural variations in grain. With hardwoods such emphasis of the grain is usually desirable. The best stains for this purpose are dyes dissolved in either water or oil. The water stains give the most pleasing result, but raise the grain of the wood and require an extra sanding operation after the stain is dry.

The most commonly used stains are the nongrained raising ones which will dry quickly. They often approach the water stains in clearness and uniformity of color. Stains on softwoods color the springwood more strongly than the summerwood, reversing the natural graduation in color in a manner that is often garish. Pigment-oil stains, which are essentially thin paints, are less subject to this objection, and are therefore more suitable for softwoods. Alternatively, the softwood may be coated with clear sealer before applying the pigment-oil stain to give more nearly uniform coloring.

In hardwoods with large pores, the pores must be filled before varnish or lacquer is applied if a smooth coating is desired. The filler may be transparent and without effect on the color of the finish, or it may be colored to contrast with the surrounding wood.

Sealer—thinned-out varnish or lacquer—is used to prevent absorption of subsequent surface coatings and prevent the bleeding of some stains and fillers into surface coatings, especially lacquer coatings. Lacquer sealers have the advantage of drying very fast.

Transparent surface coatings over the sealer may be of gloss varnish, semigloss varnish, nitrocellulose lacquer, or wax. Wax provides a characteristic sheen without forming a thick coating or greatly enhancing the natural luster of the wood. Coatings of a more resinous nature, especially lacquer and varnish, accentuate the natural luster of some hardwoods and seem to permit the observer to look down in the wood. Shellac applied by the tedious process of *French polishing* probably achieves this impression of depth most fully, but the coating is expensive and easily marred by water. Rubbing varnishes made with resins of high refractive index for light are nearly as effective as shellac. Lacquers have the advantages of drying rapidly and forming a hard surface, but require more applications than varnish to build up a lustrous coating.

Varnish and lacquer usually dry to a high-gloss surface. To reduce the gloss, the surfaces can be rubbed with pumice stone and water or polishing oil. Waterproof sandpaper and water can be used instead of pumice stone. The final sheen varies with the fineness of the powdered pumice stone, coarse powders making a dull surface and fine powders a bright sheen. For very smooth surfaces with high polish, the final rubbing is done with rottenstone and oil. Varnish and lacquer made to dry to semigloss are also available.

Flat oil finishes are currently very popular. This type of finish penetrates the wood and forms no noticeable film on the surface. Two coats of oil are usually applied, and may be followed with a paste wax. Such finishes are easily applied and maintained, but are more subject to soiling than a film-forming type of finish.

FILLING HARDWOODS

For finishing purposes, the hardwoods may be classified by the size of surface pores. Hardwoods with large pores include: ash, butternut, chestnut, elm, hackberry, hickory, Khaya (African mahogany), mahogany, oak, sugarberry, and walnut. Hardwoods with small pores include: red alder, aspen, basswood, beech, cherry, cottonwood, gum, magnolia, maple, polar, and sycamore.

Birch has pores large enough to take wood filler effectively when desired. It is still small enough as a rule, however, to be finished satisfactorily without filling.

Hardwoods with small pores may be finished with paints, enamels, and varnishes in exactly the same manner as softwoods. Hardwoods with large pores require wood filler before they can be covered smoothly with a film-forming finish. Without filler, the pores not only appear as depressions in the coating, but also become centers of surface imperfections and early failure.

UNFINISHED PANELING

Solid lumber paneling can be left unfinished in light traffic areas, on ceilings, walls, and trim not cleaned often. It's not for use in kitchens, bathrooms or other areas exposed to moisture and grease. A Danish oil can be used to give added richness to the wood, however. Apply by brushing on two coats and wiping off the excess.

Smooth-surfaced, unfinished paneling can be maintained by lightly sanding with a fine grade of sandpaper. Saw-textured unfinished paneling can be cleaned with a wire brush or coarse sandpaper. Unfinished paneling offers a completely natural appearance, especially as the wood darkens with time. Figures 6-18 through 6-20 show how to handle termite problems which may occur.

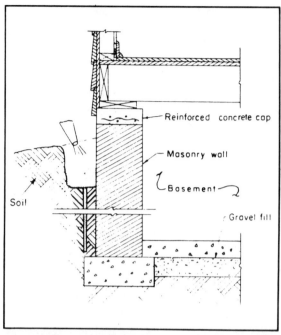

Fig. 6-19. Soil around walls and under floors of basements should be treated with insecticide.

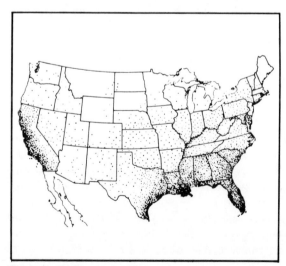

Fig. 6-18. Relative hazards of termite attack in the United States.

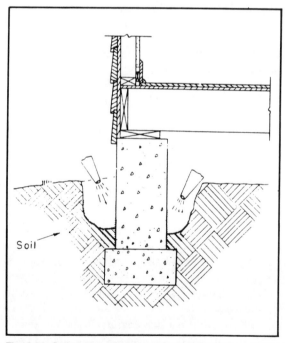

Fig. 6-20. Soil along inside and outside of foundation should also be treated with insecticide.

POPULAR FINISHES

A variety of finishes can be applied to solid lumber paneling, depending on the wood, the use, and your personal tastes. One of the more popular woods for solid lumber paneling is redwood (Table 6-1). Here are some recommended finishes for redwood and other types of solid lumber paneling:

☐ *Wax finishes*, with or without stain tint, add rich soft luster to wood and provide some water resistance. They're easier to remove if applied over two coats of sealer.

☐ *Penetrating oil*, or Danish oil, a lightly protective clear finish for walls and ceilings away from moisture, enriches wood tones and prevents wood from drying out.

☐ *Clear lacquer*, recommended in a satin texture, will protect walls, ceilings, dividers, trim, and other surfaces that are only cleaned occasionally with a dry cloth.

☐ *Clear sealers*, such as alkyd resin or polyurethane, are for all wood near heat and moisture, except in a bathroom. Two coats will darken wood, yet enhance grains and textures.

☐ *Varnishes*, semigloss alkyd resin or polyurethane, are best for kitchen/bathroom areas. Recommended two to six coats withstand hard scrubbing and give wood an appearance of depth.

☐ *Stains*, such as alkyd resin or synthetic stains, may be applied for various color effects. Wiping each coat before it dries highlights grains and textures. Wax, sealer, or varnish overcoat is recommended.

☐ *Paint* (oil base or alkyd resin) applied over alkyd resin primer forms a smooth, solid, water-resistant film, and is good for color accents on paneling and cabinet trim or edges and surfaces most touched.

To help you further in selecting an appropriate and practical finish for your solid lumber paneling, let's take a look at solid lumber panelling finishes, where to use them, how to apply them, and how to maintain them.

Wax Finishing

Finishing your solid lumber paneling with wax adds soft luster to the wood and touch-up is smooth, easy, and even. Applying wax over two coats of clear lacquer makes it easier to remove or paint over. Wax can be applied for all interior uses except kitchens and bathrooms by simply following the manufacturer's directions. On smooth-surfaced wood, use a soft cloth and rub with the wood grain. On saw-textured surfaces, apply with a stiff brush.

To maintain waxed lumber paneling, wash with mild detergent and rinse with a damp cloth. Remove grime with a nonmetallic scouring pad. Restore appearances with a new coat of wax. Wipe off the excess.

Sealers and Danish Oil

Alkyd resin and polyurethane sealers, as well as Danish oil, are clear, flat, penetrating sealers that will darken most woods, especially softwoods, appreciably. They can be used on solid lumber paneling for all interior uses except bathrooms. They are especially good for use on saw-textured surfaces.

Application of sealers and Danish oil is easy: just brush on two coats directly from the container. Maintenance is just as easy: clean with a damp cloth.

Clear Lacquer Finishing

Clear lacquer offers solid lumber paneling a natural appearance with some protection from dirt. Although it forms a film, clear lacquer isn't glossy unless many coats are applied. It will darken the wood slightly, however.

Clear lacquers can be used for walls, dividers, ceilings, trim, or anywhere that needs only dry cleaning. They should not be used in bathrooms, kitchen, or other areas requiring scrubbing. Clear lacquers can be applied by spraying (beware of open flames) or brushing on. A coat or two of wax over lacquer gives a rich luster. Buff with a soft cloth.

Clear lacquer can be cleaned with a soft cloth dampened with turpentine, mineral spirits, or water. If you've covered the lacquer with wax, follow the maintenance suggestions for wax already described.

Varnishes

Alkyd resin and polyurethane varnishes are available in flat, semiglossy or glossy textures. Varnishes seal better than lacquers and withstand hard scrubbing. They darken and deepen wood tones and may show scratches. Varnishes are recommended for all interior uses of most solid lumber paneling, especially redwood. Multiple coats are good for kitchens and bathrooms.

Varnishes can be applied by brushing on two coats for most uses or up to six coats for kitchens and bathrooms. Let dry and sand lightly between coats. You can clean varnished paneling with soapy water and a soft cloth, or you can use turpentine or mineral spirits. Scratches or nicks can be touched up with tinted wax.

Pigmented Stains

Better for interiors than paint, pigmented stains only partially obscure wood grain and texture, and are available in many colors. Pigmented stains can be used for any interior finishing. Protect from liquids, soiling, or frequent cleaning by covering the stain with a clear finish or satin sealer.

Pigmented stains can be brushed, rolled, or sprayed on. More coats mean deeper color effect. To emphasize grain or texture, apply one coat of stain and wipe the surface before it dries. A coat of thinned paint, wiped off, gives a similar effect. Maintenance is easy; just avoid heavy scrubbing that may smudge the wood.

You can improve the beauty and function of pigmented stain by applying an overcoat of a clear finish, such as lacquer, varnish, or wax. The stain then provides the color tone desired, and the overcoats provide the protection. Follow maintenance instructions for whatever finish is used as the overcoat.

APPLYING FINISHES

The common methods of applying finishes to solid lumber paneling and other surfaces are brushing, rolling, and spraying. The choice of method is based on several factors, such as speed of application, environment, type and amount of surface, type of coating to be applied, appearance of finish, and your experience. Brushing is the slowest method; rolling is much faster; and spraying is usually the fastest by far. Brushing is ideal for small surfaces and odd shapes or for cutting in corners and edges. Rolling and spraying are efficient on large, flat surfaces. Spraying can also be used for round or irregular shapes.

The general surroundings may prohibit the spraying of paint because of fire hazards or potential damage from overspraying. Adjacent areas not to be coated must be covered before you spray. This step results in loss of time and, if extensive, may offset the speed advantage of spraying.

Whereas brushing leaves brush marks after the paint is dry and rolling leaves a stippled effect, spraying yields the smoothest finish if done properly. Lacquer-type products, such as vinyls, dry rapidly and should be sprayed. Applying them by brush or roller may be difficult, especially in warm weather.

Brushes

Brushes, as with other tools, must be of high quality and maintained in perfect working condition at all times. Brushes are identified by the type of bristle used: natural, synthetic, or mixed. Nylon is the most common of the synthetic bristles. By artificially "exploding" the ends and kinking the fibers, manufacturers have increased the paint load nylon can carry and reduced the coarseness of brush marks. Other bristles include Chinese hog, horsehair, and badger hair, often preferred for varnish work.

Brushes are not only identified by their bristles, but by shapes and sizes required for specific painting and finishing jobs. (See Figs. 6-21 through 6-23.) Flat- and square-edged wall brushes range in widths from 3 to 6 inches and are used for painting large, continuous surfaces, either interior or exterior.

Sash and trim brushes are available in four shapes: flat- and square-edged, flat- and angle-

Table 6-1. Finishes for Redwood Solid Lumber Paneling.

Finish Choices	Where to Use	Application	Maintenance	Effects
Unfinished	Light traffic areas, ceilings, walls, panels and trim not cleaned often. Not for use in kitchens, bathrooms or other areas exposed to moisture and grease.	A "Danish" oil can be used to give added richness to the wood. Brush on two coats and wipe off excess.	Smooth surfaced redwood; lightly sand with fine grade sandpaper. Saw-textured redwood; clean with wire brush or coarse sandpaper.	Completely natural appearance. Wood may darken with time.
Wax	All interior uses except kitchens and bathrooms.	Follow manufacturer directions. On smooth surfaced redwood, use a soft cloth and rub with woodgrain. On saw-textured surface, apply with stiff brush.	Wash with mild detergent, rinse with damp cloth. Remove grime with nonmetalic scouring pad. Restore surface appearance with new coat of wax; wipe off excess.	Adds soft luster to redwood and touch-up is smooth, easy and even. Applying wax over two coats of clear lacquer makes it easier to remove or paint over.
Alkyd Resin & Polyurethane Sealers; Danish Oil	All interior uses except bathrooms. Good for use on saw-textured surfaces.	Two coats, brushed on.	Clean with damp cloth.	These clear, flat penetrating sealers will darken redwood appreciably.
Clear Lacquer	For ceilings, walls, dividers, trim, anywhere that needs only dry cleaning; not bathrooms, kitchens or other areas requiring scrubbing.	Apply by spraying (beware open flames), or brush on. Coat or two of wax over lacquer gives rich luster. Buff with soft cloth.	Clean with soft cloth dampened with turpentine, mineral spirits, or water. See maintenance suggestion under "Wax" for lacquer finishes topped with wax.	Natural appearance with some protection from dirt. Although it forms a film, clear lacquer isn't glossy unless many coats are applied. Will darken redwood slightly.
Alkyd Resin & Polyurethane Varnishes	All interior uses on surfaced redwood. Multiple coats good for kitchens and bathrooms.	Brush on two coats for most uses, up to six coats for kitchens and bathrooms. Let dry and sand lightly between coats.	Clean with soapy water and soft cloth. Or use turpentine or mineral spirits. Touch up scratches or nicks with tinted wax.	Available in flat, semi-gloss or glossy textures, varnishes seal better than lacquers and withstand hard scrubbing. They darken and deepen woodtones and may show scratches.
Pigmented Stains	Any interior use. Protect from liquids, soiling, or frequent cleaning by covering the stain with a clear finish or satin sealer (see "Stain and Clear Combinations").	Brush, roll on, or spray. More coats mean deeper color effect. To emphasize grain or texture, apply one coat stain, wipe surface before dry. Coat of thinned paint, wiped off, gives similar effect.	Easy maintenance. Avoid heavy scrubbing that may smudge wood.	Better for interiors than paint, stains only partially obscure redwood grain and texture, and are available in many colors.
Stain and Clear Combinations	Same as "Pigmented Stains."	Same as "Pigmented Stains." Protect from grease, dirt, liquids with one coat of wax, lacquer, or varnish.	Follow maintenance instruction for whatever finish is used as the overcoat.	Stain provides color tone desired. Overcoats provide protection.

Courtesy California Redwood Association

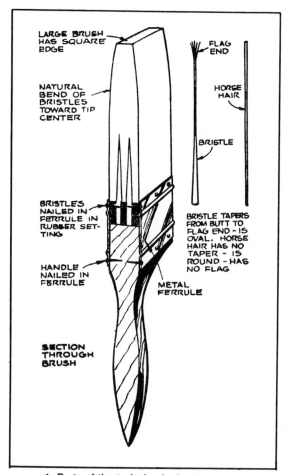

1. Parts of the typical paint brush.

Fig. 6-22. Common wide paint brush for larger surfaces.

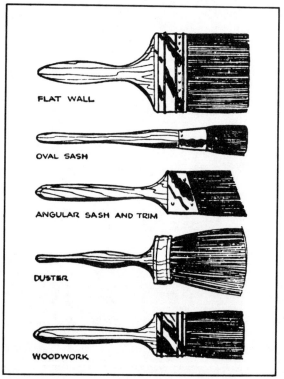

Fig. 6-23. Five types of common paint brushes.

edged, round, and oval. They range in width from 1, 1 1/2, or 3 inches or have diameters of 1/2 to 2 inches. Uses include painting window frames, sashes, narrow boards, and interior and exterior trim surfaces. For fine-line painting, the bristle end of the brush is often chisel-shaped to make precise edging easy.

Enameling and varnish brushes are flat-, square-, or chisel-edged and are available in widths from 2 to 3 inches. Their select, fine, short bristles cause relatively high-viscosity gloss finishes to lay down in a smooth, even film.

Stucco and masonry brushes are primarily for exterior wall surfaces.

Use the right size brush for the job. Avoid a

brush that is too small or too large. A large job doesn't necessarily go faster with an oversize brush. When the brush size is out of balance, you tend to apply coatings at an uneven rate; general workmanship declines; and you actually tire faster because of the extra effort required per stroke.

Synthetic bristle brushes are ready to use when purchased. The performance of natural bristle brushes is much improved by soaking 48 hours in linseed oil, followed by a thorough cleaning in mineral spirits. The bristles are made more flexible and tend to swell in the ferrule of the brush, resulting in fewer bristles working loose when the brush is used.

Rollers

A paint roller consists of a cylindrical sleeve or cover that slips into a rotatable cage to which a handle is attached. The cover may be 1 1/2 to 2 1/4 inches (inside diameter) and 3, 4, 7, or 9 inches long. Proper roller application depends on the selection of the specific fabric and the thickness of the fabric (nap length), based on the type of paint or finish used and the smoothness or roughness of the surface to be finished. The fabrics generally used for rollers are lamb's-wool pelt, mohair, Dynel and Dacron.

Lamb's-wool pelt is solvent resistant and available in nap lengths up to 1 1/4 inches. It is recommended for application of synthetic finishes on semismooth and rough surfaces. It mats badly in water and is not recommended for water-based finishes.

Mohair is made primarily of Angora hair. It is also solvent resistant. Supplied in nap lengths of 3/16 and 1/4 inch, it is recommended for application of synthetic enamels and water-based paints on smooth surfaces.

Dynel is a modified acrylic fiber that is water resistant. It is best for application of water-based paints and paints containing solvents, except strong solvents such as ketones. It is available in nap lengths ranging from 1/4 to 1 1/4 inch.

Dacron is a synthetic fiber somewhat softer than Dynel. It is best suited for exterior oil-based or latex paints. It is available in nap lengths ranging from 5/16 to 1/2 inch.

Immediately after use, rollers should be cleaned with the type of thinner recommended for the paint or finish in which the roller was used. After being cleaned with thinner, the roller should be thoroughly washed in soap and water, rinsed in clear water, and dried.

Spray Guns

A spray gun is a precision tool that mixes air under pressure with paint, breaks it up into spray, and ejects if out in a controlled pattern. There are several types, either with a container attached to the gun or with the gun connected to a separate container by means of hoses. There are bleeder or nonbleeder, external-mix or internal-mix, and pressure-feed or suction-feed guns. Figures 6-23 through 6-24 illustrate the operation of these guns.

The principal parts of the gun body assembly are shown in Fig. 6-24. The air valve controls the supply of air and is operated by the trigger. The

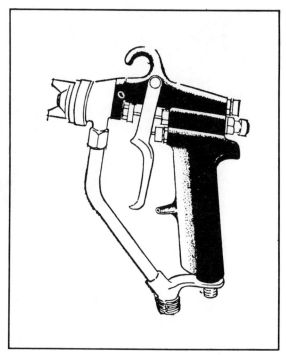

Fig. 6-24 Typical air spray painter.

129

spreader adjustment valve regulates the amount of air that is supplied to the spreader horn holes of the air cap, thus varying the paint pattern. It is fitted with a dial that can be set to give the pattern desired. The fluid needle adjustment controls the amount of spray material that passes through the gun. The spray head locking bolt locks the gun body and the movable spray head together.

In airless spray painting, the spray is created by the forcing of paint through a restricted orifice at very high pressure. The paint is atomized without the use of air jets, thus the name airless sprayer. Liquid pressures of 1500 psi and higher are developed in special air or electrically operated, high-pressure pumps and delivered to the gun through a single hose line.

The airless spray system provides a rapid means of covering large surfaces with wide-angle spray without overspray mist or rebound. The single, small-diameter hose line makes gun handling easy. The spray produced has a full, wet pattern for quick film buildup, but requires extra care in lapping and stroking to avoid excessive coverage that would result in runs, sags and wrinkles.

Brush Painting Techniques

Select the type of brush and paint pot needed for the job. The best type of pot for brush painting is a 1-gallon paint can from which the lip around the top has been removed (Fig. 6-25). (The lid of the can is fitted to the lip around the top.) You can cut this lip off with a cold chisel. When you leave the lip on the pot, it fills up with paint as you scrape the brush. This paint will flow over the lip, run down the outside of the can, and drip off.

Dip the brush to only 1/3 the length of the bristles. Scrape the surplus paint off the lower face of the brush so there will be no drip as you transfer

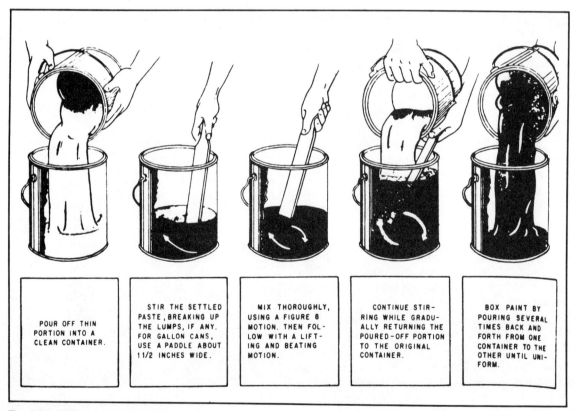

POUR OFF THIN PORTION INTO A CLEAN CONTAINER.

STIR THE SETTLED PASTE, BREAKING UP THE LUMPS, IF ANY. FOR GALLON CANS, USE A PADDLE ABOUT 1 1/2 INCHES WIDE.

MIX THOROUGHLY, USING A FIGURE 8 MOTION. THEN FOLLOW WITH A LIFTING AND BEATING MOTION.

CONTINUE STIRRING WHILE GRADUALLY RETURNING THE POURED-OFF PORTION TO THE ORIGINAL CONTAINER.

BOX PAINT BY POURING SEVERAL TIMES BACK AND FORTH FROM ONE CONTAINER TO THE OTHER UNTIL UNIFORM.

Fig. 6-25. Mixing paint manually.

the brush from the pot to the surface to be painted. Refer to Figs. 6-26 through 6-29.

For complete coverage in applying the finish by brush, use the laying-on, laying-off method. Use long, horizontal brush strokes first (laying-on), then cross your first strokes by working up and down (laying-off). By laying-on and laying-off, you distribute the paint evenly over the surface, cover the surface completely, and use a minimum amount of paint. A good rule is to lay on the paint the shortest distance across the area and lay off the longest distance. When applying finish to walls or any vertical surface, you should lay on in horizontal strokes and lay off in vertical strokes.

If appropriate, always paint the ceiling first, working from the far corner. By doing so, you can keep the wall free from drips by wiping as you go along. Be sure to carry a rag for wiping up. You will also find that finish coats on the ceiling should normally be laid-on for the shortest ceiling distance and laid-off for the longest ceiling distance.

To avoid brush marks when finishing up a square, you should use strokes directed toward the last square finished, gradually lifting the brush near the end of the stroke while the brush is still in motion. Every time the brush touches the painted surface at the start of a stroke, it leaves a mark. For this reason, you should never finish a square by

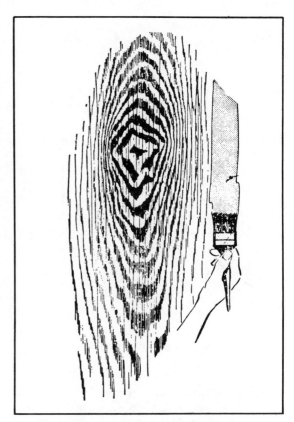

Fig. 6-27. Laying paint or stain with the grain.

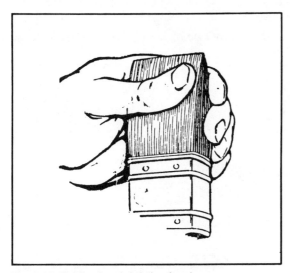

Fig. 6-26. Testing brush bristles for shape.

Fig. 6-28. Cover in broad sweeps of the brush.

Fig. 6-29. Pay special attention to coverings knots.

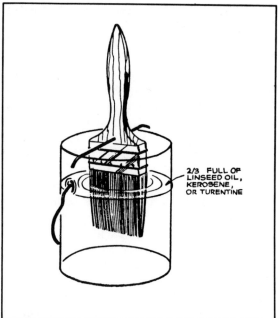

Fig. 6-31. A brush can be stored for a short time in a can of linseed oil, kerosene, or turpentine.

2/3 FULL OF LINSEED OIL, KEROSENE, OR TURENTINE

brushing toward the unpainted area, but always end up by brushing back toward the area already painted. Refer to Figs. 6-30 through 6-32.

Roller Painting Techniques

Pour premixed paint into a tray to about 1/2 its depth. Immerse the roller, then move it back and forth along the ramp of the tray to fill the cover completely and remove excess paint. As an alternative to using the tray, place a specially designed, galvanized-wire screen into a 5-gallon can of finish. This screen attaches to the can and remains at the correct angle for loading and spreading paint on the roller.

To remove entrapped air from the roller cover, work the first load of paint out on newspaper. You

Fig. 6-30. A brush can be combed after cleanup to separate bristles.

Fig. 6-32. If no other method is available, you can store a brush by wrapping it tightly.

are now ready to apply the finish. You should have already trimmed around corners, moldings, etc. In rolling paint or other finish onto a surface, always work a dry area into the just-painted area. Never roll completely in the same or one direction. Don't roll too fast to avoid spinning the roller at the end of a stroke. Always feather out final strokes to pick up any excess finish on the surface. Do this by rolling out the final stroke with minimal pressure. Get as close as possible to surfaces already trimmed by brush, maintaining the same texture.

Spray Painting Techniques

When you squeeze the trigger of a spray gun, an air valve opens to admit compressed air. The air passes through the gun body to the spray head. In an external-mix type of spray head, the air doesn't come in contact with the paint inside the gun, but is blown out through small holes drilled in the air cap. A thin jet of paint is shot out of the nozzle, and the force of the air striking it spreads the jet into a fine spray.

Complete instructions for the care, maintenance, and operation of a spray gun are contained in the manufacturer's manual, and these instructions should be carefully followed. The handling of a spray gun is best learned by practice, but here are some helpful tips for using a spray gun to finish solid lumber paneling.

Before starting to spray, check adjustments and operation of the gun by testing the spray on a surface similar to that which you intend to coat. There are no set rules for spray gun pressure or distance to hold the gun from the surface because pressure and distance vary considerably with the nozzle, the paint used, and the surface to be coated. The minimum pressure necessary to do the work is the most desirable, and the distance is normally from 6 to 10 inches.

Always keep the gun perpendicular to and at the same distance from the surface being painted (Figs. 6-33 and 6-34). Start the stroke before squeezing the trigger, and release the trigger before completing the stroke. When the gun is not held perpendicular or is held too far away, part of the

paint spray will evaporate and strike the surface in a nearly dry state. This is called *dusting*. When you fail to start the stroke before starting the spray or spraying to the end of the stroke, it will cause the finish to build up at each end of the stroke and will run or sag. When you arch the stroke, it is impossible to deposit the paint in a uniform coat.

When spraying an inside or outside corner, stop 1 or 2 inches short of the corner, as shown in Fig. 6-35. Do so on both sides. Then turn your gun on its side and, starting at the top, spray downward coating both sides at once.

In spraying a large area into which small parts and pieces protrude, first coat them lightly, then go over the whole surface. In painting an office, for example, first spray the doorframes and all small items secured to the walls, then the entire office. Following this method eliminates a lot of touching up later.

FINISHING UP

In this chapter you've learned how to finish up solid

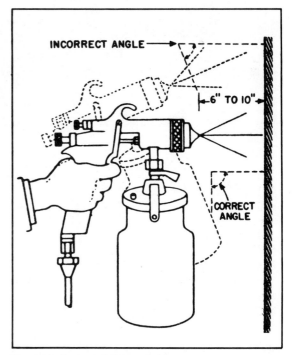

Fig. 6-33. Spray painting techniques.

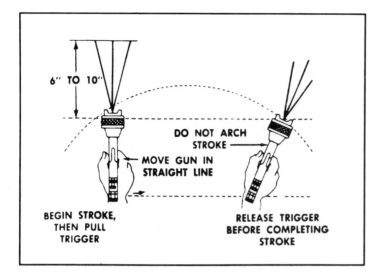

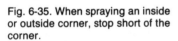

Fig. 6-34. Edging with a spray painter.

6" TO 10"

DO NOT ARCH STROKE

MOVE GUN IN STRAIGHT LINE

BEGIN STROKE, THEN PULL TRIGGER

RELEASE TRIGGER BEFORE COMPLETING STROKE

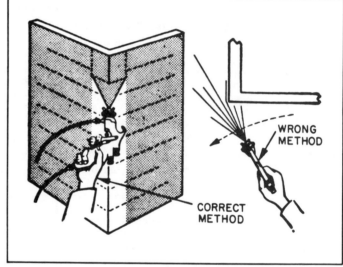

Fig. 6-35. When spraying an inside or outside corner, stop short of the corner.

WRONG METHOD

CORRECT METHOD

lumber paneling using a wide variety of finishes and equipment. You are now ready to tackle a few projects on your own. The next two chapters include both easy and fancy solid lumber paneling projects that will guide you in applying what you've learned in the first six chapters of this book. Have fun!

Chapter 7

Easy Solid Lumber Paneling Projects

BY NOW YOU SHOULD BE ANXIOUS TO GET started on your first solid lumber paneling project. Thus far you've learned how to plan paneling projects, select solid lumber paneling, prepare paneling, choose and use fasteners and adhesives, and install and finish solid lumber paneling.

In this chapter you'll discover numerous solid lumber paneling projects that you can complete with this basic knowledge and no prior skills. Some projects will be very obvious, while others may be new applications to you of solid lumber paneling. All will be easy enough to be planned and completed by a beginner.

Before you begin, however, browse through the first few chapters of this book again. Look especially at the illustrations there. Refresh your memory on how to select and install solid lumber paneling. Make notes if you wish. Plan to enjoy your first solid lumber paneling project.

CEDAR CLOSET

One of the easiest solid lumber paneling projects is the installation of tongue-and-groove cedar panels to the inside of an existing closet, thus making a cedar closet. The sweet smell of aromatic red cedar, though pleasant to humans, repels destructive moths. For this reason, storing clothes and woolens in a cedar closet offers maximum protection. Cedar aroma doesn't kill moths, but it keeps moths out of the closet. Out-of-season woolens and furs should be dry-cleaned before storage. This kills all moth larvae.

Lining an existing closet with cedar provides year-round moth protection for valuable garments and no home should be without at least one cedar-lined closet. A project of this kind can be accomplished in a single weekend by the average home handyman using everyday tools such as hammer, saw, and nails. Figures 7-1 through 7-8 will help you install an aromatic red cedar closet lining as your first solid lumber paneling project.

Red cedar for lining closets comes in strip form. The red cedar boards are 3/8 inch thick, and 1 to 8 feet long. It is manufactured in several face

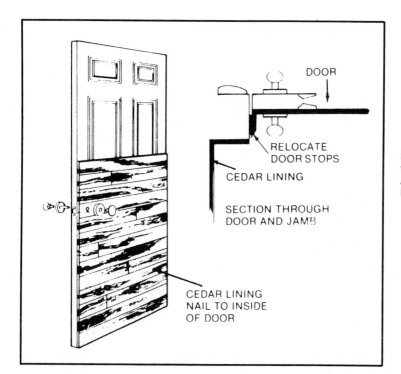

Fig. 7-1. Lining the door of a closet for an easy cedar closet (courtesy Aromatic Red Cedar Closet Lining Manufacturers Association).

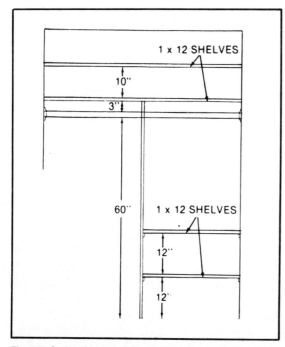

Fig. 7-2. Cedar closet shelving (courtesy Aromatic Red Cedar Closet Lining Manufacturers Association).

widths from 2 1/2 to 4 1/2 inches. Cedar lining, available in home centers and lumberyards, comes in bundles or cartons, in random or uniform lengths. Individual pieces are tongued-and-grooved along edges and ends to simplify construction.

Measure the closet to determine square footage needed, including the ceiling, floor, and inside of door. Remove door stops and base moldings and locate wall studs, usually at the points where moldings were nailed to the wall. Snap a chalk line on the wall to indicate the nailing line along each stud.

Apply cedar lining horizontally to the existing wall (Fig. 7-3), beginning at the floor with the groove side down. Then build up, row by row, until the wall is covered. Tap boards down lightly with a hammer to ensure a snug fit before nailing. Only one nail is needed to secure the individual cedar pieces to each stud. Use small finishing nails or wire brads. The interlocking of the cedar's tongue-and-groove edges ensures a tight, sturdy fit. If the boards are uneven at the corners, use pine quarter-round molding to cover the gaps. Remember to

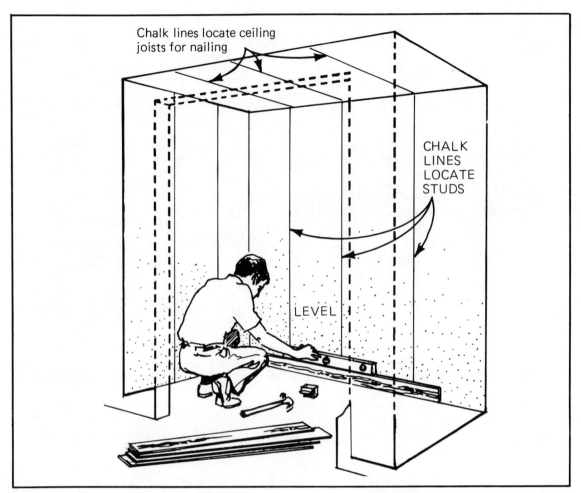

Fig. 7-3. Installing the first horizontal cedar panel (courtesy Aromatic Red Cedar Closet Lining Manufacturers Association).

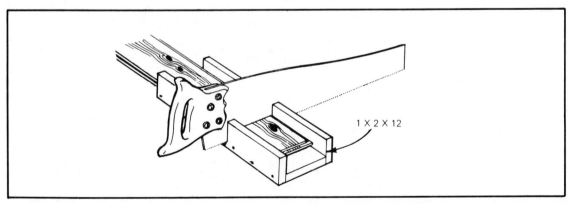

Fig. 7-4. Using a homemade saw guide to cut cedar panels (courtesy Aromatic Red Cedar Closet Lining Manufacturers Association).

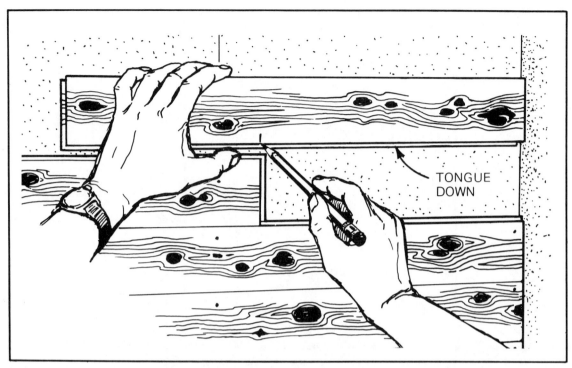

Fig. 7-5. Marking a cedar panel before installation (courtesy Aromatic Red Cedar Closet Lining Manufacturers Association).

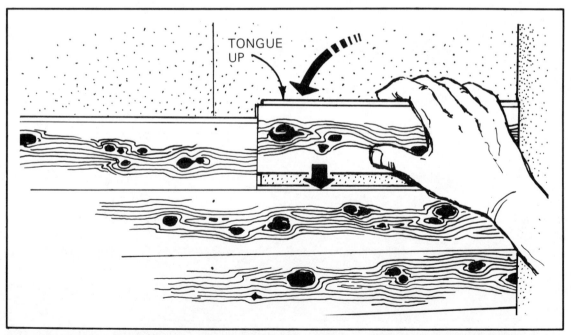

Fig. 7-6. Installing cut cedar panel (courtesy Aromatic Red Cedar Closet Lining Manufacturers Association).

Fig. 7-7. Installing the final panels to your cedar closet (courtesy Aromatic Red Cedar Closet Lining Manufacturers Association).

Fig. 7-8. Installing cedar panels on the floor of your cedar closet (courtesy Aromatic Red Cedar Closet Lining Manufacturers Association).

stain the molding to look like the cedar before installation. After lining the closet, add weather stripping around the door to make the closet as airtight as possible.

Wipe cedar occasionally with a dry cloth to remove dust which may clog pores. Never use varnish, shellac, or other finish. Doing so would seal in the cedar aroma. If the aroma fades slightly over the years, rub the surface lightly with fine sandpaper or steel wool to open the pores and renew the cedar fragrance.

OUTDOOR PLANTERS

Planters are basically boxes—short or tall, simple or fancy, formal or rustic—and they are all meant to hold soil so you can grow decorative plants or vegetables where you wish. Many planters can be made with solid lumber paneling (Fig. 7-9 through 7-13).

You affect the appearance of the planter by the shape you choose and the wood you work with. Solid lumber panels, 2-inch boards, 3/4-inch exterior grade plywood, or plywood siding can all be used. Pressure-treated fir or pine lumber or redwood is recommended for soil-contact situations, but other materials will do if you line the planter with polyethylene film. Wood may be treated with a preservative like pentachlorophenol, but such materials may be toxic to plants for a variable period of time, so wait 2 to 3 weeks before planting. The plastic liner is a good way to go in all cases, even if you work with naturally resistant species like redwood.

A basic approach to planter construction is to first cut the sides and join them with waterproof

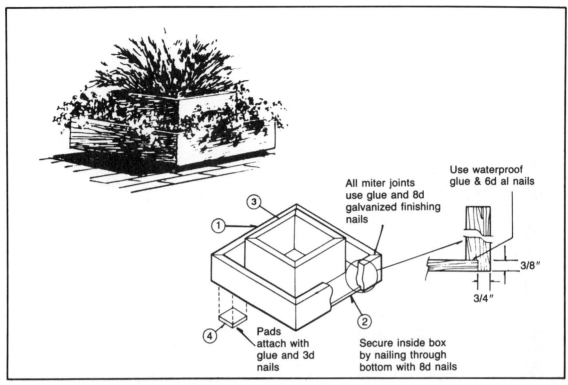

All miter joints
use glue and 8d
galvanized finishing
nails

Use waterproof
glue & 6d al nails

3/8"

3/4"

Pads
attach with
glue and 3d
nails

Secure inside box
by nailing through
bottom with 8d nails

Fig. 7-9. Easy-to-build solid lumber paneling planter (courtesy Georgia-Pacific).

Fig. 7-10. Large planter (courtesy Georgia-Pacific).

Fig. 7-11. Legged planter of solid lumber panels (courtesy Georgia-Pacific).

Fig. 7-12. Planters can be varied in size (courtesy Georgia-Pacific).

glue and galvanized nails. Then add an inside frame of 1-×-1-inch lumber to support a tongue-and-groove solid lumber panel bottom. Drill 3/8-inch drain holes about 6 inch inches apart. Cover the holes with insect screening. Add 1-×-1 corner blocks to the inside using glue and nails. Add a top frame of 1- or 2-inch lumber to give the planter a finished look. The final step is to cut pieces of 1 × 4 or 2 × 4, 4 inches long to use as feet. Then you can easily move the container.

SIMPLE SHELVING

There are many ways to build shelves using solid lumber paneling. There are few areas in a home

Fig. 7-13. Matching triangular planters of solid lumber panels (courtesy Georgia-Pacific).

that can't be made more usable or more attractive through the use of shelving. Shelving can be pretty or not, but they must always be practical. The installation area and the purpose of the shelves should be important guides. An easy-to-do storage shelf in a garage for storing paint cans and such can simply be ready-made metal brackets that screw to studs to hold a section of tongue-and-groove solid lumber panels.

To a great extent, what you will place on the shelves will determine the spacing between shelves and even the type of material you should use. Average books can be stored on shelves that are spaced 9 to 10 inches, but there are those coffee table volumes which need more height. Allowing for off-sizes at the start—by varying shelf spacing or by including a two-shelf-high nook—is a good idea.

A few books aren't heavy, but a 10-foot row has

considerable weight. Decide beforehand whether to use 2-inch stock for extra strength over long spans or 1-inch boards with the added support of vertical dividers between shelves.

The depth of shelves should not be arbitrary. Narrow shelves that do the job are better than excessively wide ones since they take up less space, are cheaper to install, and minimize areas that must be dusted. Often, shelves of various widths provide the answer to a problem. If so, use the wider

shelves at the bottom areas of the project.

Solid lumber paneling makes good shelving whether the shelves are built-in or merely spanning ready-made brackets. Solid lumber panels can be constructed from narrower boards to make wider shelving. A good depth is 9 1/2 to 10 inches.

Try to anticipate future needs when you install nonadjustable, built-in shelving. More shelf space than you need right now is better than too little tomorrow. Figures 7-14 through 7-16 offer ideas for

Fig. 7-14. Solid Lumber panel shelving (courtesy Georgia-Pacific).

Fig. 7-15. Wall shelves from solid lumber panels (courtesy Georgia-Pacific).

building shelves from solid lumber paneling. Look at other homes and in home magazines to come up with other ideas for making practical shelving from solid lumber paneling.

FLOWER BOXES

Here are a couple of flower boxes that you can easily make with 1-×-6 or 1-×-8 solid lumber paneling for an entry hall or other interior location. Fig-

Fig. 7-16. Divider from solid lumber panels (courtesy Georgia-Pacific).

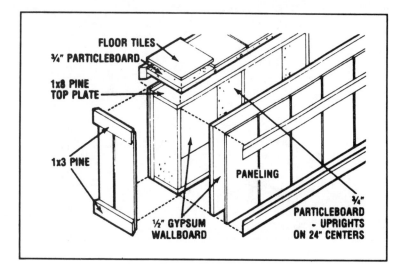

ure 7-17 illustrates the construction of a flower box and stand built from a base made out of a solid piece of wood 1 1/4 inches thick. At each corner a block 1 1/4 × 2 3/4 × 2 inches is attached. A middle support 1 1/4 × 2 3/4 × 6 inches is attached along one side. The top piece of the base, the same size and thickness as the bottom, is attached over these supports. The box itself is made out of 1-inch stock for sides, top, and bottom.

Use butt joints on all these pieces. The box is centered on the base and attached to it with wood screws. The screws should be counter sunk. The inside of the box can be lined with either zinc or copper. All seams should be soldered so that moisture from the earth inside the box will not reach the wood.

To build this simple flower box and stand you will need the following materials:

2 pieces	1 1/4″ × 9″ × 37″
1 piece	1 1/4″ × 6″ × 2 3/4″
4 pieces	1 1/4″ × 2″ × 2 3/4″
2 pieces	1″ × 6″ × 36″
2 pieces	1″ × 6″ × 6″
1 piece	1″ × 6″ × 34″

Copper or zinc lining

The second flower box is illustrated in Fig. 7-18. It can easily be constructed to stand alone or mount on a wall covered with solid lumber paneling in your hall, bedroom, living room, or bathroom. The box is made by building a simple rectangular box 36 inches long, 8 inches wide, and 8 inches deep. Use 1-×-8 stock for the sides and bottom and 1/4-inch wood for the ends. A strip of 1 × 1 5/8 inches is attached flush with the bottom and another flush with the top. Midpoint between these two, a third strip of 1-×-1 5/8-inch stock fits in place. Line the inside of the box with copper or zinc.

Here are the materials you will need to build this flower box:

2 pieces	1″ × 8″ × 34″
1 piece	1″ × 6″ × 34 1/2″
3 pieces	1″ × 2″ × 36″
6 pieces	1″ × 2″ × 6 3/4″
2 pieces	6 3/4″ × 7 5/8″ × 1/4″

STORM DOOR

There are dozens of applications of solid lumber paneling other than as a wall treatment. Here's one

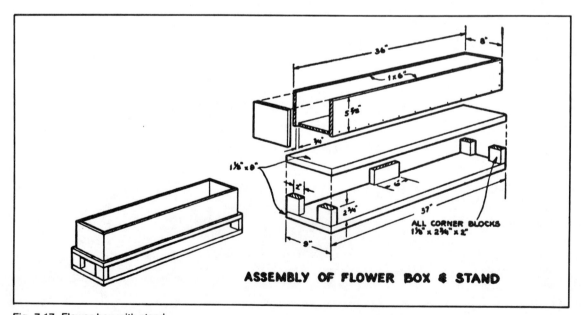

ASSEMBLY OF FLOWER BOX & STAND

Fig. 7-17. Flower box with stand.

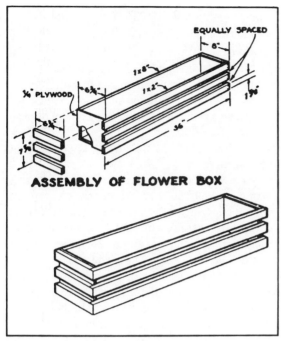

Fig. 7-18. Flower box without stand.

of them: a storm door (Fig. 7-19).

To construct your solid lumber paneling storm door, select only well-seasoned 6-inch tongue-and-groove boards, or random width boards, and keep them in a warm, dry place until the door has been assembled and painted. If exposed to damp weather, the boards will expand and later shrink, resulting in a poor fitting job.

The dimensions for the door in Fig. 7-19 are 3 feet (36 inches) × 6 feet 8 inches, but they can be varied to suit individual requirements. Measure the door opening carefully and use the actual dimensions.

The assembly of the door begins with removing the groove from one of the boards with either a plane or saw. This board will be the first one on the hinge side of the door. All other boards are left with tongue and groove intact, except the final board on the latch side of the door which will have the tongue removed. If it is necessary to decrease the width of a board considerably to get a good fit, it is better to take an equal amount off both the first and last boards rather than all of it off one.

Refer to Fig. 7-20. The boards are held in place by means of two 1-×-6-inch cleats. The top cleat should have its lower edge 5 feet 9 inches from the bottom of the door. The lower cleat should be 12 inches from the bottom. Bevel the outside edges of the cleats and use flathead wood screws to fasten the boards. Be sure that all joints between boards are driven up tight and that the ends of the boards are level. As soon as all the boards are in place, cut a diagonal brace out of 3/4-×-4-inch stock to run between the cleats. The ends of this brace should fit snugly up against the cleats, and it should be fastened to the door boards with screws.

The 8-×-12-inch window opening in the door is made 5 feet from the bottom and positioned in the exact center of the door. Start the opening with a brace and bit, making a hole at each corner. Connect the holes by cutting first with a keyhole saw and then, when the cut is large enough, with a crosscut saw. The slide for the glass consists of two pieces of 1/4-×-3/4-inch wood fastened 1/4 inch above and below the window opening. Attached

Fig. 7-19. Storm door built from lumber panels.

145

over these is a strip of 1/4 × 1 inch brought flush with the opening. The window glass slides in the 1/4-inch recess at top and bottom. Use double-strength or plate glass.

Here are the materials you will need to build the storm door as illustrated:

6 pieces	6″ tongue-and-groove boards, 6′ 8″
1 piece	3/4″ × 4″ × 6′
2 pieces	1″ × 6″ × 24″
2 pieces	1/4″ × 1″ 1′ 11″
2 pieces	1/4″ × 1″ × 10″
2 pieces	1/4″ × 3/4″ × 1′ 11″
2 pieces	1/4″ × 3/4″ × 10″
1 piece glass	8 1/2″ × 12 1/2″ (double-strength or plate with finger grip)
2 butt strap hinges	13″

SHUTTERS

To go with your solid lumber storm door, here are four shutters that you can easily construct from solid lumber paneling. Well designed and well-constructed shutters can do much toward improving the outside appearance of a house. Besides adding a decorative note, they can make undersize windows appear larger. The four shutters shown in Figs. 7-21 through 7-24 cover a wide range of architectural styling.

Since shutters are exposed to the weather, it is best to make them out of woods, such as cypress or redwood, which are highly resistant to decay. Noncorrosive screws should be used in the construction of the shutters and to fasten the hinges in place. The blind holdbacks used to hold the shutters open should also be of noncorrosive metals because any rusting at this point will leave unsightly stains on the house siding. As noted in the illustrations, each shutter is fastened to the window trim

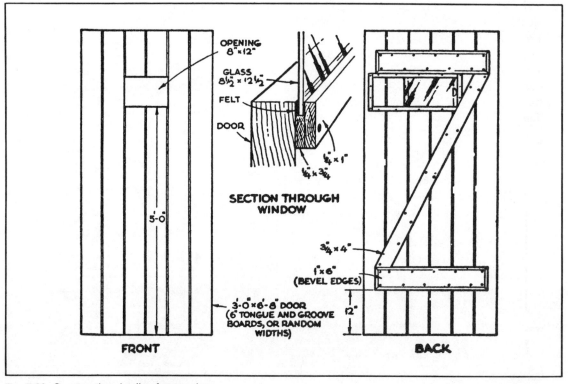

Fig. 7-20. Construction details of storm door.

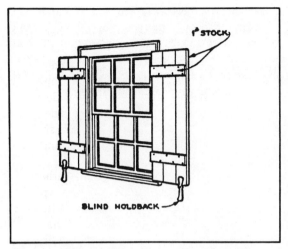

Fig. 7-21. Batten shutters from solid lumber panels.

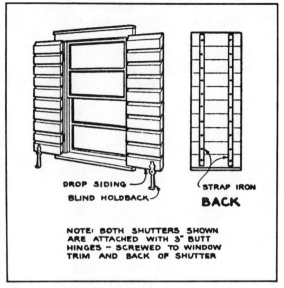

NOTE: BOTH SHUTTERS SHOWN ARE ATTACHED WITH 3" BUTT HINGES – SCREWED TO WINDOW TRIM AND BACK OF SHUTTER

Fig. 7-23. Drop siding shutters.

with 3-inch butt hinges.

Shutter No. 1 (Fig. 7-21) is a simple batten type suitable for almost any house. One-inch wide stock is used in the construction. The width will depend on the size of the window opening. Try as far as possible to have all the upright pieces more or less the same width.

Shutter No. 2 (Fig. 7-22) uses tongue-and-groove boards. It is fastened together in the back of the shutter with lengths of strap iron.

Shutter No. 3 (Fig. 7-23) uses the same construction as No. 2. Here, though, drop or novelty

siding is used in place of the tongue-and-groove boards of the second shutter.

Shutter No. 4 (Fig. 7-24) is made of 3/4-inch exterior boards. It has 1/4- × -1-inch strips nailed diagonally across the face.

Shutters should be painted before installation to avoid paint dripping on the house siding.

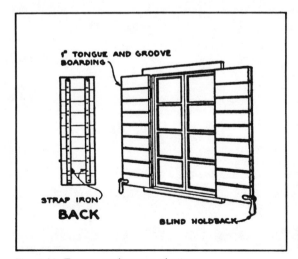

Fig. 7-22. Tongue-and-groove shutters.

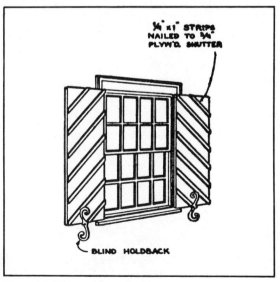

Fig. 7-24. Diagonal face shutters.

EASY PANELING

These are just a few of the many easy projects that can be constructed with solid lumber paneling. As you estimate the amount of solid lumber paneling you need to cover your living room, bedroom, bath, or hallway wall, add a few extra board feet and make some of these easy projects to match.

Chapter 8 will discuss some fancier solid lumber projects to dress up your home. It will both provide projects and get you started thinking about all of the many applications of solid lumber paneling throughout your home and business.

Chapter 8

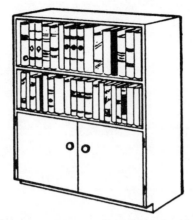

Fancy Solid Lumber Paneling Projects

N OW THAT YOU'VE LEARNED THE BASICS OF selecting and installing solid lumber paneling and you've tried a few easy projects, it's time to get fancy. In this final chapter you'll learn how to be even more creative in the installation of all types of solid lumber paneling throughout your home.

Fancy solid lumber paneling projects may require that you have either completed some easy projects or had some related project experience, such as remodeling or light construction. If so, you will find both practicality and enjoyment in completing the more complex projects of this chapter. So grab your hammer and a square, and we'll begin some fancy solid lumber paneling projects.

ATTIC CEDAR CLOSET

Converting wasted space into a mothproof cedar storage closet can give new life to unused attic space. The procedure is simple. After you've selected the site, frame it out with 2- × -4 lumber on 16-inch centers, as in Fig. 8-1. If a part of your new closet uses existing plaster walls or ceiling, locate the studs behind them by driving experimental nails into the surfaces. Nail holes are no problem because they will be covered by the cedar lining.

Start with an inside corner and, working up from the floor from left to right, place cedar boards with the groove side down. Only one nail is needed to secure the individual cedar pieces to each stud. Use small finishing nails or wire brads set about 1/2 inch from the edge. The interlocking of the cedar's tongue-and-groove edges ensures a tight, sturdy fit regardless of whether or not the joint occurs over a stud. If the boards are uneven at the corners, use pine quarter-round molding to cover the gaps. Stain the molding to look like cedar before you install it.

Once the walls are finished, line the floor and ceiling using the same procedure (Fig. 8-2). The door should also be lined to ensure the most complete moth protection possible. Install weather stripping around the doorframe to make the closet as airtight as possible. See Fig. 8-3.

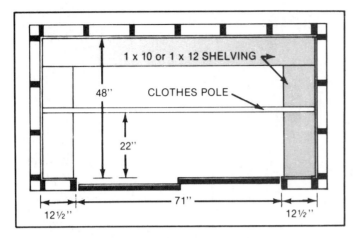

Fig. 8-1. Details of attic cedar closet (courtesy Aromatic Red Cedar Closet Lining Manufacturers Association).

1 x 10 or 1 x 12 SHELVING

CLOTHES POLE

48"

22"

71"

12½"

12½"

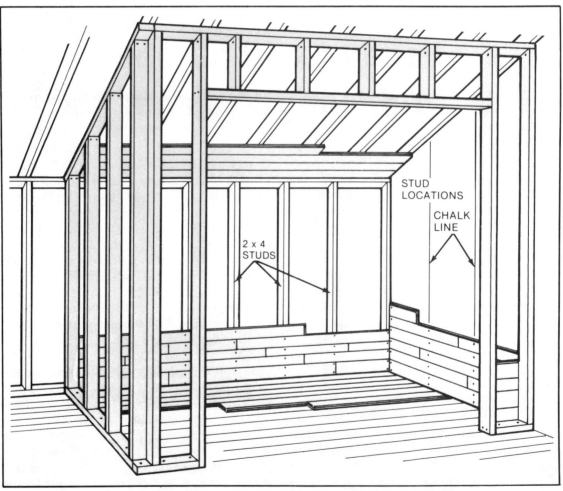

STUD LOCATIONS

CHALK LINE

2 x 4 STUDS

Fig. 8-2. Lining the floor (courtesy Aromatic Red Cedar Closet Lining Manufacturers Association).

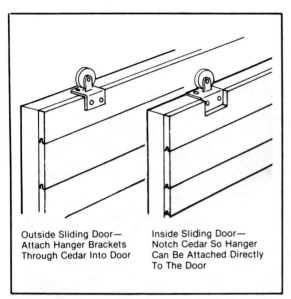

Outside Sliding Door—
Attach Hanger Brackets
Through Cedar Into Door

Inside Sliding Door—
Notch Cedar So Hanger
Can Be Attached Directly
To The Door

Fig. 8-3. Cedar door detail (courtesy Aromatic Red Cedar
Closet Lining Manufacturers Association).

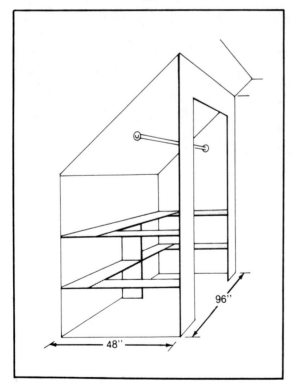

96"

48"

Fig. 8-4. Hanger rod and shelving detail (courtesy Aromatic
Red Cedar Closet Lining Manufacturers Association).

To determine hanger rod and shelving design, consider what is to be stored. In most storage areas, 4 1/2 feet will be adequate for overcoats and dresses. Most men's suits require only 3 1/2 feet. Hanger rods can be adjusted accordingly (Fig. 8-4). As you can see, an attic cedar closet is a useful addition to your home.

FREE-STANDING CEDAR CLOSET

If you need more storage space for your clothes, consider building a free-standing cedar chest (Fig. 8-5). Because the closet shown is portable, it makes a wise investment for both homeowners and apartment dwellers. The time to build a vertical cedar closet is before you find moth holes in your clothes.

Here's how to build your free-standing cedar closet:

☐ All framing is made from 1- x -3 pine or poplar. Corner joints should be dowelled and glued.

☐ Make the top and bottom of the unit from 3/4-inch plywood or 1-inch cedar shelving boards glued edge-to-edge to make the 22-inch width. Shelves are made of the same cedar boards.

☐ Line each frame with tongue-and-groove cedar strips. Start with the back panel first. Draw a border line 3/4 inch in from the edge on all four sides. This line represents the overlap of sides, top, and bottom (Fig. 8-6).

☐ Cut the first cedar strip at a 45-degree angle. Fit it in place across the frame starting at the border line. Since cedar strips are tongued-and-grooved on the ends, full-length strips are not necessary.

☐ Accurately glue and nail the first strip in place. Following strips fit into the first strip and are glued and tacked in place as you go. Allow the cedar strips to overlap the border lines slightly. When all of the strips are in place, set your electric saw to a depth of 3/8 inch and trim all cedar ends at once.

☐ Attach cedar strips to side panels similarly on a 45-degree angle. Cedar strips should be flush with the front edge and rabbet line at top and bottom. At the back edge allow a 3/8-inch setback to clear the cedar strips on the back panel.

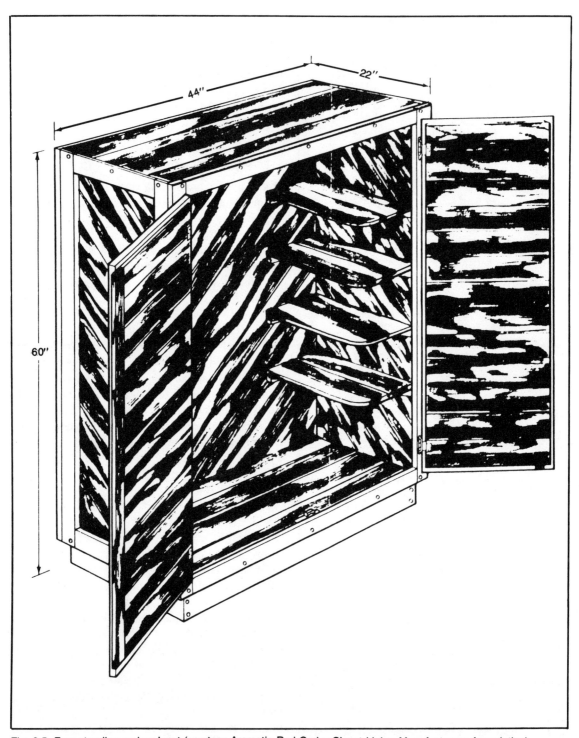

Fig. 8-5. Free-standing cedar closet (courtesy Aromatic Red Cedar Closet Lining Manufacturers Association).

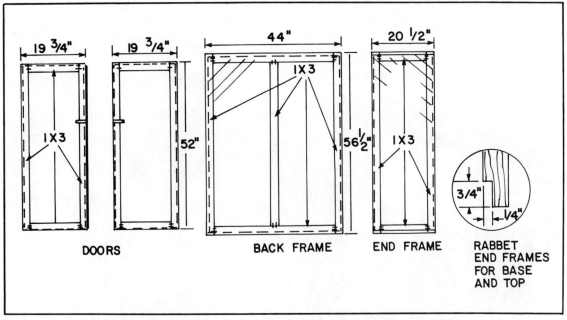

Fig. 8-6. Closet sections (courtesy Aromatic Red Cedar Closet Lining Manufacturers Association).

☐ On the inside of the door frames mark parallel border lines 3/8 inch in from the hinge edge, top, and bottom. Glue and nail a 3/8-×-1-inch strip on the left-hand door so that it forms a 3/8-inch lip where the doors meet (Fig. 8-7). On the right-hand door mark a 3/8-inch parallel border line to compensate for the lip of the left-hand door.

☐ Cut cedar strips to length to fit crosswise between border line. Glue and nail in place.

☐ Make the base from 1-×-4 boards fastened with screws at the corners. Add 45-degree diagonal brackets at the corners. Attach the base to the bottom panel through brackets at each corner.

☐ Assemble the completed panels to each other with glue and 1 1/2-inch-×-#10 wood screws, as shown in Fig. 8-8. Countersink screws 1/4 inch below the surface and plug holes with red cedar plugs.

☐ Fasten the top and bottom to the panels with glue and wood screws, as before.

☐ Attach hinges to the doors and fit in place. Fasten the left-hand door to the frame with screws. Then position the right-hand door so that the space between the doors is about 1/16 inch and fasten it in place with screws.

☐ Glue and screw 1-×-2 shelf cleats for 1-×-10 shelves. Shelves should be fastened to cleats with three 1 1/2-inch-×-#8 wood screws each. Add the hanger pole with brackets.

☐ Make a handle from a 1-inch diameter dowel. Flatten the back of the dowel where it attaches to the door frame and fasten it in place with screws.

☐ Add a 3/4-×-1-inch cleat and magnetic catch at the bottom of right-hand door. If needed, add a cleat and catch at the top also.

☐ Varnish the outside only with high-gloss varnish clear acrylic. Do not varnish the inside.

Your free-standing cedar chest is now complete and ready to use.

BATHROOM PARTITION

In scouting your house for solid lumber paneling projects, you may decide to redecorate the bathroom with paneling. That's great, but don't stop there. You can also install a partition screen and

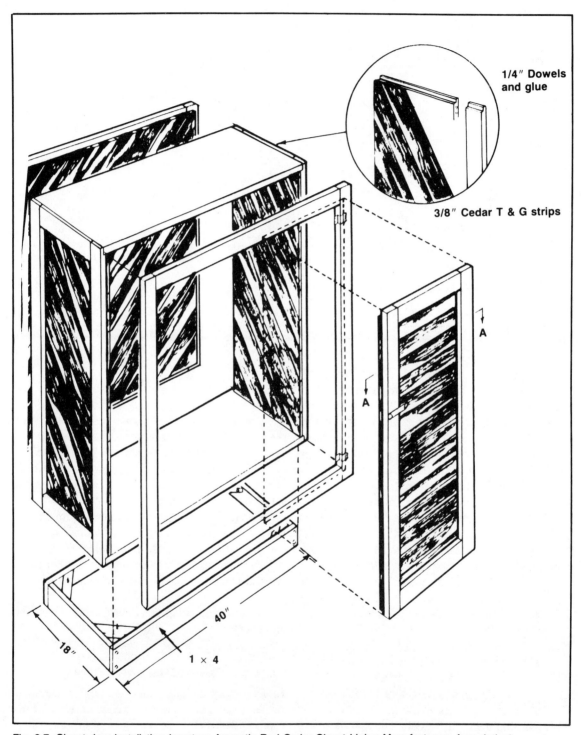

1/4″ Dowels and glue

3/8″ Cedar T & G strips

A

A

40″

18″

1 × 4

Fig. 8-7. Closet door installation (courtesy Aromatic Red Cedar Closet Lining Manufacturers Association).

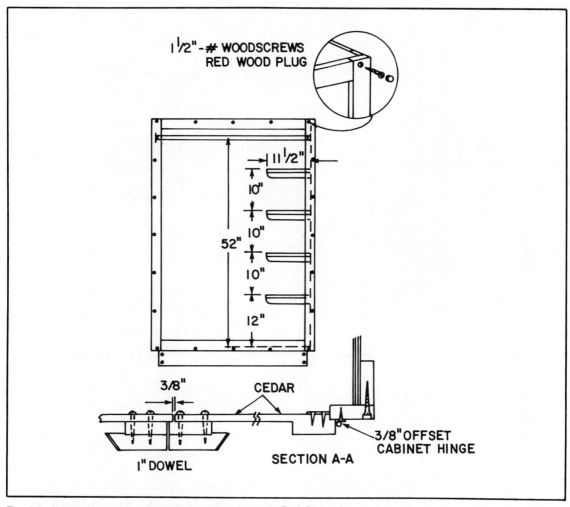

Fig. 8-8. Assemble completed panels (courtesy Aromatic Red Cedar Closet Lining Manufacturers Association).

cover it with solid lumber paneling (Figs. 8-9 through 8-12).

A partition screen can make sense in many areas—a block between a toilet and sink area, a "wall" to create a nook for a dressing table or desk, a projection into a room as a backing for a chair or even a bed. It can be designed as a room accent or finished to blend with existing walls. It's a quick and simple construction project which requires only 2-×-4 framing (or 2-×-2 framing if you wish to save space) and a covering of solid lumber paneling.

Secure the sole plate by nailing into the floor, and the top plate by nailing into joists. Use a plumb bob to be sure the plates are correctly aligned. Add the studs—three full-length ones for a solid wall— or additional framing if you wish to have a see-through section or a mirror.

For a solid partition, use inside corner molding for the edge at the existing wall and outside corner molding for the open end. Base molding is installed at the floor, and cove molding at the ceiling edge.

SAUNA LOOK

Whether you decide to install a sauna in your bath-

Fig. 8-9. Bathroom partition covered with solid lumber paneling (courtesy Georgia-Pacific).

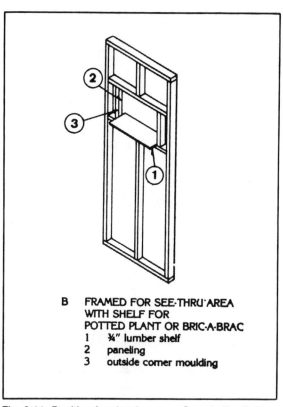

B FRAMED FOR SEE-THRU AREA
WITH SHELF FOR
POTTED PLANT OR BRIC-A-BRAC
1   ¾" lumber shelf
2   paneling
3   outside corner moulding

Fig. 8-11. Partition framing (courtesy Georgia-Pacific).

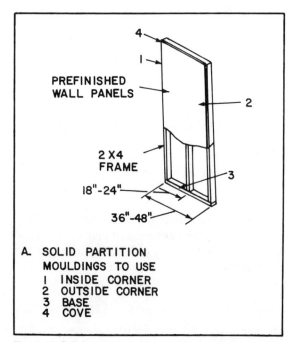

PREFINISHED
WALL PANELS

2 X 4
FRAME

18"-24"

36"-48"

A. SOLID PARTITION
MOULDINGS TO USE
1   INSIDE CORNER
2   OUTSIDE CORNER
3   BASE
4   COVE

Fig. 8-10. Solid partition details (courtesy Georgia-Pacific).

room or not, you can add the sauna look to it. It's accomplished by covering walls and ceiling with redwood or cedar exterior or interior paneling. The idea is to provide a total wood look with standard materials that are suitable for a moisture area.

If existing walls are sound and smooth, you may be able to attach new materials directly by nailing into studs. For uneven walls, cover the surface with a system of 1- × -2 furring strips, as suggested earlier in this book. They can be shimmed out where necessary to compensate for wall irregularities. Use standard wood moldings to cover corners and joints at the floor and the ceiling. Replace frames and doors of cabinets with the same material used on the walls.

STORAGE WALLS

Here's a great idea! While you're planning on covering a wall in your living room, family room, or den

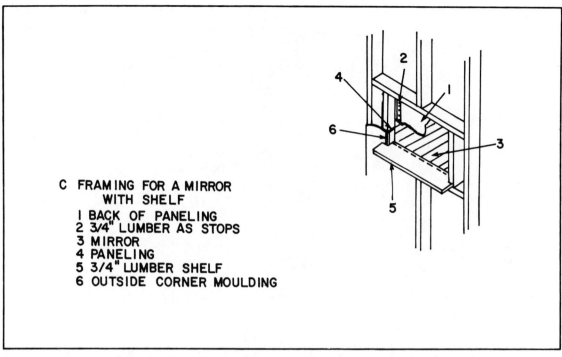

C FRAMING FOR A MIRROR
        WITH SHELF
   1 BACK OF PANELING
   2 3/4" LUMBER AS STOPS
   3 MIRROR
   4 PANELING
   5 3/4" LUMBER SHELF
   6 OUTSIDE CORNER MOULDING

Fig. 8-12. Framing for mirror with shelf (courtesy Georgia-Pacific).

with solid lumber paneling, plan to install shelves and make it a storage wall. Wall storage units are usually thought of as fixed shelves and dividers. They may serve an initial purpose, but they can become a static bore. What's the solution?

Consider making modular units that you can rearrange at will. An example is a set of 1-×-6-foot cases with two or three shelves. The cases can be used vertically, horizontally, or by combining the two positions. The project is most economical if the cases are sized in relation to standard widths of solid lumber, i.e., 1 × 12. Make rabbet cuts at corners and use dadoes for the shelves. Assemble with glue and finishing nails.

Providing a storage wall this way does require more material than the usual shelves-on-brackets design. The extra flexibility, however, justifies the additional cost.

HOME BAR

There's no such thing as a stock bar, especially if one is custom-designed for home use. The traditional counter/stool arrangement can run parallel to a wall or storage unit or it can turn to enclose a corner (Figs. 8-13 and 8-14). Assuming stools are 2 1/2 feet high, the corner should be about 3 1/2 feet tall. Projecting the counter top about 6 inches will provide knee room and area for a bar rail or a foot-support step. A bar built in as a divider must be sized to suit the space.

Actually, the concept is more for the mixer's convenience. Minimum project space is required because drinks are toted to other seating areas. If space is at a premium, design the bar as a movable cart. It can be completely closed with hinged doors and a swing-up top to resemble any cabinet when not in use.

A general construction approach is to view the project as a skeleton made with 2-×-3 kiln-dried pine or fir covered with solid lumber paneling. A decorative top can be made with tongue-and-groove solid lumber panels that are edge-glued and edge-nailed, then finished naturally.

Fig. 8-13. Beautiful bar decorated with solid lumber paneling (courtesy Georgia-Pacific).

Fig. 8-14. Small bar that can fold up (courtesy Georgia-Pacific).

ATTIC STUDIO

An attic remodeling offers the opportunity to provide areas that are exactly right for special activities. With a few tools, an attic can be remodeled and paneled to become and artist's studio. A skylight is especially useful if installed on the north side. Existing rafters can be used as supports for framed skylights you make yourself. Use 2-×-2 pressure-treated lumber for frames, grooving inside edges to receive double-strength glass. You might consider using Lexan, which is a new, shatterproof plastic. Be sure to use plenty of caulking or to do a conventional putty job when you install the glass.

Mount the frames between rafters on well-caulked and nailed stops made of 1-×-1 pressure-treated lumber.

Another way to go is to check out skylight domes which are preassembled and ready for installation. Refer to my book *Doors, Windows and Skylights* (TAB book No. 1578).

Adding a knee wall beneath the skylight will improve inside appearance. The space behind the wall will be usable if you leave openings for hinged doors. Help to brighten the area even more by covering walls and ceiling with light-toned solid lumber paneling.

PICNIC TABLE

Figure 8-15 illustrates a fancy picnic table that you can construct with solid lumber paneling, as well as dimensional lumber. Slabs for the table and the benches can be 1-×-6 or 2-×-4 pine, fir, or redwood attached with brass screws to 2-×-4 cleats. The cleats keep the boards together, but are also a way to attach understructures. The most simple legs are H-shaped, attached to the lumber cleat with 1/4-inch carriage bolts and braced with a 2-×-4 stretcher between the horizontal members.

You can achieve a different look if you make the legs like open-end boxes using 2 × 4s spanned with pieces of solid lumber paneling. Size the boxes to fit over the lumber cleats and attach them there with glue and flathead screws. The design for both table and benches can be the same. Make the bench slab 10 to 14 inches wide and about 17 inches high.

Fig. 8-15. Picnic table of solid lumber panels and dimension lumber (courtesy Georgia-Pacific).

The table should be at least 36 inches wide and about 29 inches high. It's length is at least 5 1/2 feet.

Many picnic tables are made with attached benches. Run a 2-×-4 or 2-×-6 rail across the legs as a support for 1-×-6 or 2-×-12 seats. Bench tops can also be made with 2 × 4s covered with diagonal pieces of solid lumber paneling.

For a novel idea, leave an opening in the table so you can inset a copper-lined planter box. You can make it out of scrap solid lumber paneling.

BOOK CABINET

A handy combination such as this bookshelf and cabinet (Fig. 8-16) makes a very useful piece of furniture for small apartments, bedrooms, or studies. With the exception of the hardboard backing, 1-inch solid lumber paneling is used throughout the construction (Fig. 8-17).

The bottom of the cabinet should be cut 8 1/4 × 34 1/2 inches. Along the bottom, 1 1/4 inches in from the outside edge, fasten a strip of 1-×-2 lumber on edge. This piece should be cut 34 1/2

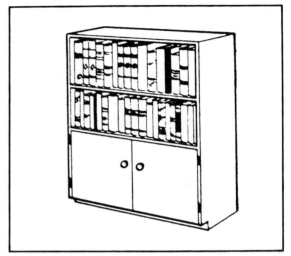

Fig. 8-16. Bookshelf and cabinet combination.

inches long. Two side pieces should be 42 inches high and 9 inches wide and made of one, two, or more solid lumber panels. On the lower outside corner of each piece, cut a notch 2-×-2 inches. The two shelves are supported at the ends by 1/2-inch quarter-round cleats fastened to the side pieces.

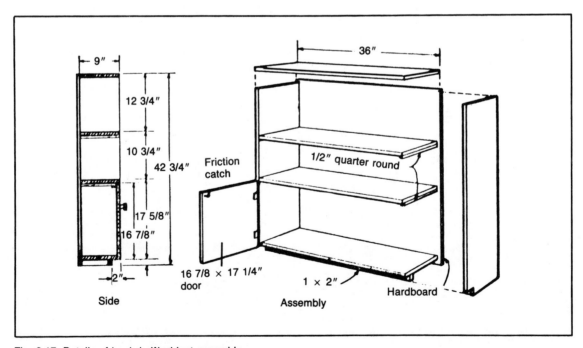

Fig. 8-17. Details of bookshelf/cabinet assembly.

The cleats for the first shelf should be 8 1/4 inches long and those for the top 9 inches long.

When the cleats have been installed, assemble the bottom and side pieces and the two shelves. The shelves should each be 9 inches wide and 34 1/4 inches long. The top of the cabinet, which is 9- × -36 inches, is now fastened in place as is the 1/8-inch hardboard back, which measures 42 3/4 × 36 inches.

The cabinet doors are 16 7/8 × 17 1/4 inches. They should be hinged with flush door hinges. Each door is fitted with a friction catch and a stock door pull.

Here is the materials list the the solid lumber paneling book cabinet:

2 pieces	1″ × 9″ × 42″
1 piece	1″ × 9″ × 36″
2 piece	1″ × 9″ × 34 1/2″
1 piece	1″ × 8 1/4″ × 34 1/2″
1 piece	1″ × 2″ × 34 1/2″
1 piece	1/8″ hardboard 42 3/4″ × 36″
1/2″ quarter round 34 1/2″	

MAGAZINE STORAGE RACK

Figures 8-18 through 8-20 illustrate an attractive magazine storage rack which can be constructed with solid lumber paneling by the more serious furniture builder. Much of the stock is 1 × 6 and 1 × 12 which can be made up of 1 × 3 tongue-and-groove solid lumber paneling.

Each shelf is 50 1/4 × 11 5/8 inches. Fasten a strip of 1 × 2 along the front ends and sides of the bottom shelf to keep it off the floor. Make up the end piece 1 × 12 × 29 1/4 inches. The back is made out of 1/8-inch hardboard 30 × 50 1/4 inches. Assemble these three pieces.

Nine inches in from the end of the lower shelf install a support for the second shelf. This support should be 8 1/2 inches high cut from 1 × 12. Put the second shelf in place, use a level to see that it is true, and then nail it to the support. Nail it also to the side piece by driving nails through the side into the end grain of the shelf. Seams where the side piece and support join the shelf can be covered

Fig. 8-18. Magazine storage rack of solid lumber panels.

with strips of 1/2-inch quarter round.

The other shelves and supports are installed similarly. The second support is the same size as the first; the third is reduced in height to 8 3/8 inches. The top shelf must be 51 inches long instead of 50 1/4 inches as it rests on the side piece.

The magazine rack is made of 1 × 6. Top and bottom are 51 inches long; side pieces are 26 3/4 inches. The backing is made of 1/8-inch hardboard 28 1/4 × 50 1/4 inches. Exactly 13 3/4 inches down from the top of the rack, install a shelf 1 × 6 × 49 1/2 inches. The back of the racks is made of 1/2-inch plywood fastened to 1-×-2-inch and 1-×-1-inch beveled cleats. To prevent magazines from slipping off, 1/2-inch quarter round should be tacked at the edge of each shelf.

Here are the materials you will need to build your magazine storage rack. The remainder are given in Figs. 8-18 through 8-20.

1″ × 12″ × 21′ 5 7/8″
1″ × 6″ × 17′ 1″
1″ × 2″ × 13′ 4 5/8″
1/2″ quarter round × 16′ 6″

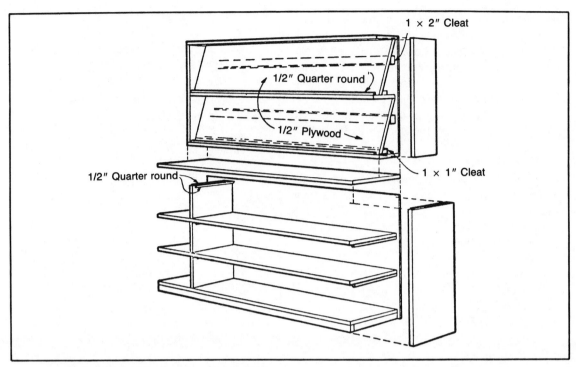

Fig. 8-19. Details of magazine storage rack assembly.

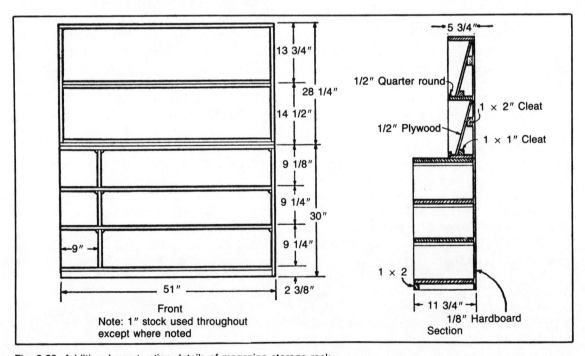

13 3/4"

28 1/4"

14 1/2"

9 1/8"

9 1/4"

30"

9 1/4"

9"

51"

2 3/8"

Front

Note: 1" stock used throughout
except where noted

5 3/4"

1/2" Quarter round

1 × 2" Cleat

1/2" Plywood

1 × 1" Cleat

1 × 2

11 3/4"

1/8" Hardboard

Section

Fig. 8-20. Additional construction details of magazine storage rack.

162

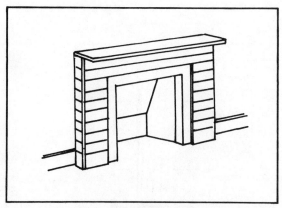

Fig. 8-21. Old-fashioned fireplace mantle.

FIREPLACE MANTLE

Many an otherwise attractive living room have been ruined by an unsightly fireplace mantle (Fig. 8-21). For those who enjoy working with wood, this situation is easily fixed.

Remove the old mantle and trim. Install nailing bases as shown in Fig. 8-22, Section A-A, and Fig. 8-23, Plan at B-B. The 1- × -2 and 2- × -2 stock are used for this purpose. Fasten these bases to the fireplace masonry with expansion bolts and to the plaster wall with toggle bolts. Before you start on the mantle, check the dimensions of the fireplace, since they may vary somewhat from those given in the illustrations.

The mantle is assembled as a unit and then moved into place and secured to the bases provided around the fireplace. The only exception to this is the pine board mantle shelf, which is attached directly to the base over the opening.

The outside corners of the mantle assembly are held together with a strip of 1- × -1 stock running up on the inside of the corner. Screws are run through this piece into the 4-inch V-joint pine boards. A 3/4-inch quarter-round molding is used to cover up the outside ends of the boards, as shown in Plan B-B. Be sure to leave a 4-inch clearance between the woodwork and the fireplace opening on all three sides. Allowing the woodwork to come any closer to the opening is a fire hazard. When the mantle has been assembled, place it in position and secure it to the nailing bases with finishing nails. The nailheads should be countersunk and the holes filled with a wood filler.

Getting a good fit along the top edge between the mantle shelf and the V-joint pine boards may

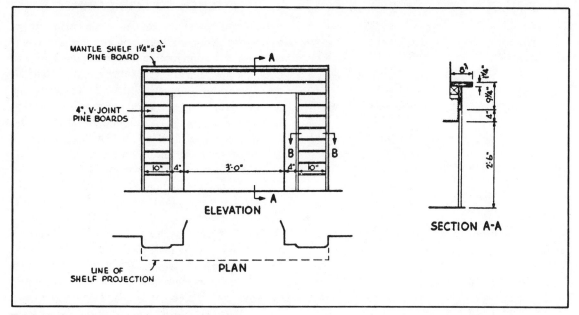

Fig. 8-22. Construction details of fireplace mantle.

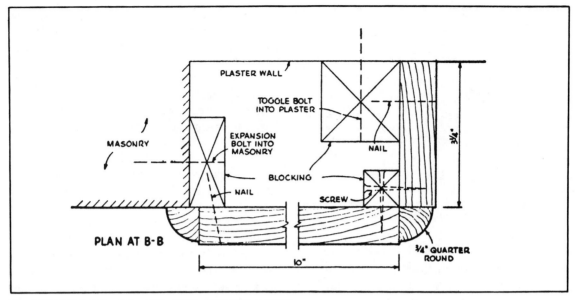

Fig. 8-23. Additional construction details of fireplace mantle.

be a problem if the fireplace or floor is slightly out of plumb. This can be corrected by trimming down the top of the mantle assembly until it fits snugly.

Here are the materials you will need:

4 pieces	2″ × 2″ × 3′ 6″
2 pieces	1″ × 2″ × 3′ 6″
1 piece	1″ × 1″ × 5′ 4″
2 pieces	1″ × 1″ × 3′ 6″
1 mantle shelf	1 1/4″ × 8″ × 5′ 5 1/2″
2 pieces	3/4″ quarter round × 38″
2 pieces	3/4″ quarter round × 30″
2 V-joint boards	4″ × 5′ 4″
14 V-joint boards	4″ × 10″
14 V-joint boards	4″ × 3 1/4″

For convenience in buying, here are the materials totals for your fireplace mantle:

2″ × 2″ × 14′
1″ × 2″ × 7′
1″ × 1″ × 12′ 4″

3/4″ quarter round × 11′ 4″
4″ V-joint boards × 25′ 10 3/4″
Remainder as just itemized

KITCHEN CORNER CABINET

Solid lumber paneling can be used in various ways to decorate your kitchen. You can build or cover cabinets with solid lumber paneling. Here's how to build and install a kitchen corner cabinet. Used either alone or in conjunction with base cabinets, this corner cabinet uses space ordinarily wasted in most kitchens. Figures 8-24 through 8-26 illustrate the construction and installation of a kitchen corner cabinet using solid lumber paneling.

Cut out the two shelves from 1/2-inch plywood 35 × 35 inches, as indicated in the plan. They will be useful as guides for making up the framework. The three rear uprights are made 2 × 2 × 35 1/4 inches. The two uprights in the front that serve as a frame for the door are 2 × 2 × 34 1/2 inches. These two pieces rest on a piece of 1 × 6 notched out as shown in the detail. Uprights are fastened together with a piece of 1 × 3. The lower pieces of 1 × 3 are set 3/4 inch up from the ends of the

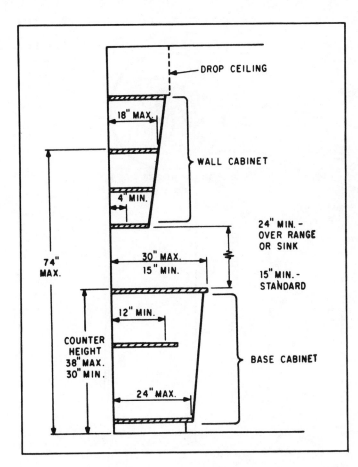

Fig. 8-24. Construction of typical kitchen cabinets.

DROP CEILING

18" MAX.

WALL CABINET

4" MIN.

24" MIN. -
OVER RANGE
OR SINK

30" MAX.
15" MIN.

15" MIN. -
STANDARD

74"
MAX.

12" MIN.

COUNTER
HEIGHT
38" MAX.
30" MIN.

BASE CABINET

24" MAX.

Fig. 8-25. Kitchen corner cabinet.

rear three uprights and flush with the ends of the two front uprights that rest on the 1 × 6. A piece of 1/2-inch plywood 3 3/8 × 14 with the ends beveled to 45-degree angles should be installed between the front uprights at the base.

Place the lower shelf in position and fasten it to the 1-×-3-inch strips and to the uprights. Eighteen inches from the bottom shelf, install four more pieces of 1 × 3 for the second shelf. Along the back of the cabinet, fasten two pieces of 1 × 8, one 35 1/8 inches long and the other 35 7/8 inches long. These pieces should extend above the top of the uprights 6 3/4 inches. Notch out the lower portions of the ends so they come flush with the edges of the 2 × 2 uprights. The top of the cabinet, which is 3/4-inch plywood or solid lumber paneling 35 1/8 × 35 1/8 inches, is then cut to size and fastened in place.

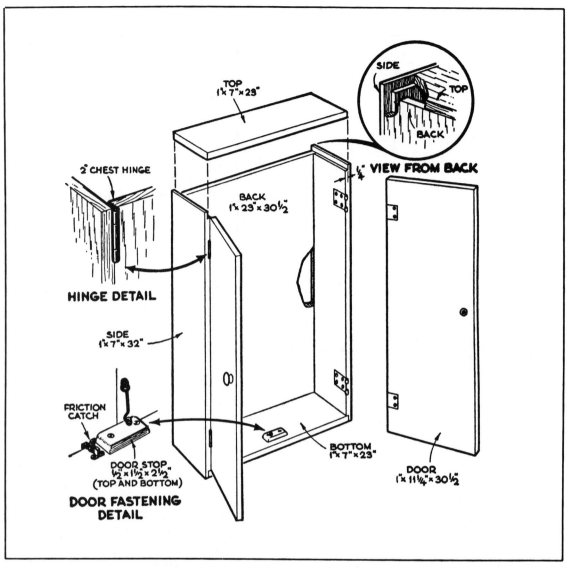

Fig. 8-26. Construction details of kitchen corner cabinet.

The latch side of the cabinet is made of 3/8-inch plywood or solid lumber paneling 24 3/8 × 35 1/4. The panel on this side should be beveled to a 45-degree angle. The hinged side of the cabinet is made of 3/8-inch plywood or solid lumber paneling 24 1/4 × 35 1/4 inches.

The cabinet door is made out of 3/4-inch plywood or solid lumber paneling. The hinge side of the door is beveled to 45 degrees, and the door is hinge to the upright with two 3- × -3-inch hinges. Two 1- × -2-inch beveled door stops are fastened to the upright on the latch side of the door, and a cabinet spring catch is also installed. The top of the counter is covered with linoleum or waterproofed solid lumber paneling in the same manner as for the base cabinet.

Here is a list of the materials you will need to build your kitchen corner cabinet:

3 pieces	2″ × 2″ × 35 1/4″	2 pieces 1/2″ plywood	35″ × 35″ or equivalent solid lumber paneling
2 pieces	2″ × 2″ × 34 1/2″		
1 piece	1″ × 8″ × 35 7/8″		
1 piece	1″ × 8″ × 35 1/8″	1 piece 1/2″ plywood	3 3/8″ × 14″ or equivalent solid lumber paneling
1 piece	1″ × 6″ × 26 3/4″		
2 pieces	1″ × 3″ × 35″		
2 pieces	1″ × 3″ × 34 1/4″	1 piece 3/8″ plywood	24 3/8″ × 35 1/4″ or equivalent solid lumber paneling
6 pieces	1″ × 3″ × 23 1/4″		
2 pieces	1″ × 2″ × 2″		
1 piece 3/4″ plywood	35 1/8″ × 35 1/8″, or equivalent solid lumber paneling	1 piece 3/8″ plywood	24 1/4″ × 35 1/4″ or equivalent solid lumber paneling
1 piece 3/4″ plywood	15 1/4″ × 31 3/4″ or equivalent solid lumber paneling	1 piece linoleum	35 1/8″ × 35 1/8″ or equivalent solid lumber paneling
Metal counter trim	25′ 1 7/8″	1 door pull	
1 pair hinges	3″ × 3″	1 friction latch	

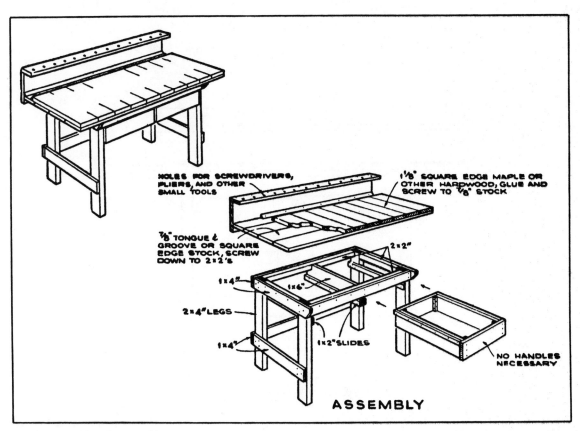

Fig. 8-27. Workbench assembly.

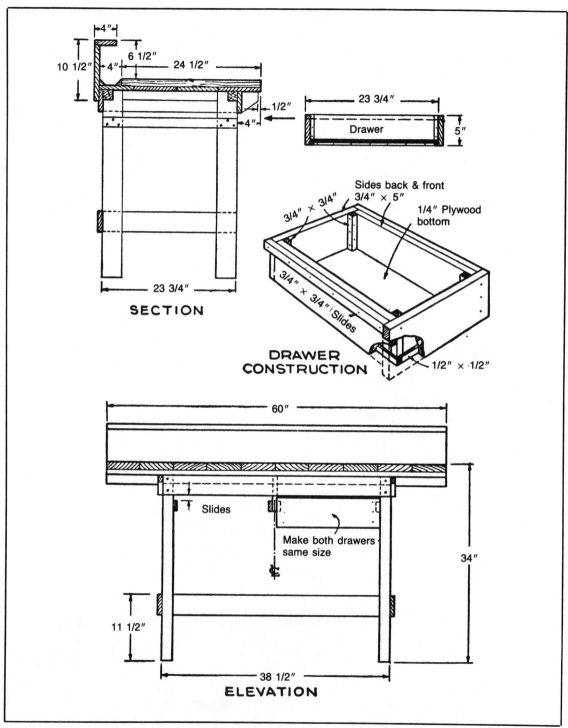

Fig. 8-28. Workbench details.

GETTING FANCIER

There are numerous home improvement projects where solid lumber paneling can be used. You can use solid lumber paneling to make cabinets, shelves, workbench (Fig. 8-27 and 8-28), furniture, walls, ceilings, toys, and even tools. Solid lumber paneling projects are only limited by your imagination.

As you browse through your local hardware, lumber, and home improvement stores, keep an eye out for other easy and fancy projects you can build with solid lumber paneling. You'll also find ideas in popular magazines such as *Mechanix Illustrated*, *The Handyman*, *1,00l Home Ideas* and even *Mother Earth News*.

Another good source of solid lumber paneling projects and assistance is your local building materials retailer. Many have free copies of plans they have tried which can make your home more beautiful and functional with solid lumber paneling.

Glossary

air-dried lumber—Lumber that has been piled in yards or sheds for any length of time. For the United States as a whole, the minimum moisture content of thoroughly air-dried lumber is 12 to 15 percent, and the average is somewhat higher. In the South, air-dried lumber may be no lower than 19 percent.

anchor bolts—Bolts to secure a wooden sill plate to concrete or masonry floor or wall.

apron—The flat member of the inside trim of a window placed against the wall immediately beneath the stool.

backband—A simple molding sometimes used around the outer edge of plain rectangular casing as a decorative feature.

base or **baseboard**—A board placed against the wall around a room and next to the floor to properly finish the space between the floor and plaster or paneling.

base molding—Molding used to trim the upper edge of interior baseboard.

base shoe—Molding used next to the floor on interior baseboard; sometimes called a carpet strip.

batten—Narrow strips of wood used to cover joints or as decorative vertical members over plywood or wide boards (Fig. G-1).

beam—A structural member transversely supporting a load.

bearing partition—A partition that supports any vertical load in addition to its weight.

bearing wall—A wall that supports any vertical load in addition to its own weight.

blind nailing—Nailing in such a way that the nailheads are not visible on the surface of the work—usually at the tongue of matched boards.

boiled linseed oil—Linseed oil in which enough lead, manganese, or cobalt salts have been incorporated to make the oil harden more rapidly when spread in thin coatings.

brace—An inclined piece of framing lumber applied to the wall or floor to stiffen the structure; often used on walls as temporary bracing until

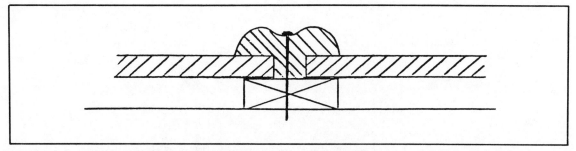

Fig. G-1. Batten (courtesy Hardwood Plywood Manufacturers Association).

framing has been completed.

butt joint—The junction where the ends of two timbers or other members meet in a square-cut joint (Fig. G-2).

casing—Molding of various widths and thicknesses used to trim door and window openings at the jambs.

column—In architecture, a perpendicular supporting member, circular or rectangular in section, usually consisting of a base, shaft, and capital. In engineering, a vertical structural compression member which supports loads acting in the direction of its longitudinal axis.

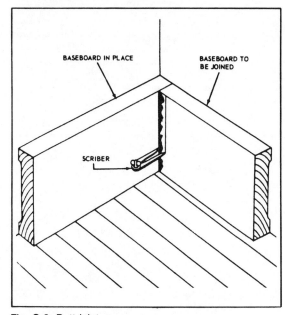

Fig. G-2. Butt joint.

condensation—In a building, beads or drops of water (and frequently frost in extremely cold weather) that accumulate on the inside of the exterior covering of a building when warm, moisture-laden air from the interior reaches a point where the temperature no longer permits the air to sustain the moisture it holds. A vapor barrier under the gypsum lath or drywall on exposed walls will reduce condensation in them.

construction, drywall—A type of construction in which the interior wall finish is applied in a dry condition, generally in the form of sheet materials or wood paneling, as contrasted to plaster.

construction, frame—A type of construction in which the structural parts are wood or depend upon a wood frame for support (Fig. G-3). In codes, if masonry veneer is applied to the exterior walls, the classification of this type of construction is usually unchanged.

corner bead—A strip of formed sheet metal, sometimes combined with a strip of metal lath, placed on corners before plastering to reinforce them. Also, a strip of wood finish, quarter-round or angular, placed over a plastered corner for protection.

cove molding—A molding with a concave face used as trim or to finish interior corners.

crown molding—A molding used on cornice or wherever an interior angle is to be covered.

d—See *Penny*.

dado—A rectangular groove across the width of a board or plank. In interior decoration, a special type of wall treatment.

171

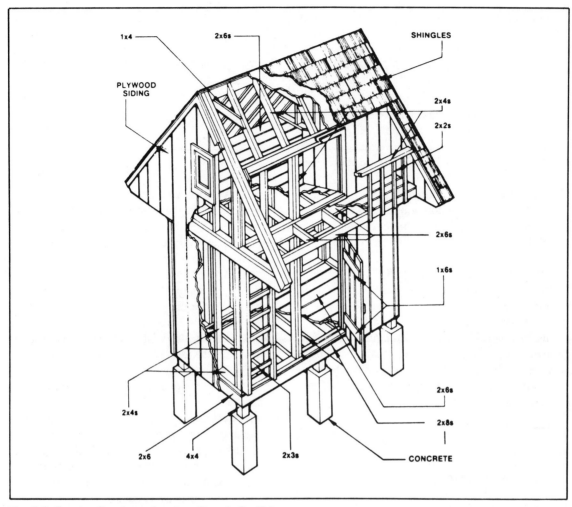

Fig. G-3. Construction, frame (courtesy Georgia-Pacific).

decay—Disintegration of wood or other substance through the action of fungi.

direct nailing—To nail perpendicular to the initial surface or to the junction of the pieces joined; also termed *face nailing*.

dormer—An opening in a sloping roof, the framing of which projects out to form a vertical wall suitable for windows or other openings.

dressed and matched (tongued and grooved)—Boards or planks machined in such a manner that there is a groove on one edge and a corresponding tongue on the other.

drywall—Interior covering material, such as gyp-

sum board or plywood, which is applied in large sheets or panels.

filler (wood)—A heavily pigmented preparation used for filling and leveling off the pores in open-pored woods.

fire-retardant chemical—A chemical or preparation of chemicals used to reduce flammability or to retard the spread of flames.

flat paint—An interior paint that contains a high proportion of pigment and dries to a flat or lusterless finish.

framing, balloon—A system of framing a

building in which all vertical structural elements of the bearing walls and partitions consist of single pieces extending from the top of the foundation sill plate to the roof plate and to which all floor joists are fastened.

framing, platform—A system of framing a building in which floor joists of each story rest on the top plates of the story below or on the foundation sill for the first story, and the bearing walls and partitions rest on the subflooring of each story.

furring—Strips of wood or metal applied to a wall or other surface to even it. It normally serves as a fastening base for finish material, such as plywood paneling or solid lumber paneling.

gloss enamel—A finish material made of varnish and sufficient pigments to provide opacity and color, but little or no pigment of low opacity. Such an enamel forms a hard coating with maximum smoothness of surface and a high degree of gloss.

grain—The direction, size, arrangement, appearance, or quality of the fibers in wood (Fig. G-4).

Fig. G-4. Grain.

grain, edge (vertical)—Lumber that has been sawed parallel to the pith of the log and approximately at right angles to the growth rings, i.e., the rings form an angle of less than 45 degrees or more with the surface of the piece.

grain, flat—Lumber that has been sawed parallel to the pith of the log and approximately tangent to the growth rings, i.e., the rings form an angle of less than 45 degrees with the surface of the piece.

grain, quartersawn—Another term for edge grain.

gusset—A flat wood, plywood, or similar type member used to provide a connection at the intersection of wood members. Most commonly used at joints of wood trusses, they are fastened by nails, screws, bolts, or adhesives.

heartwood—The wood extending from the pith to the sapwood, the cells of which no longer participate in the life processes of the tree.

insulation board, rigid—A structural building board made of coarse wood or cane fiber in 1/2-inch and 25/32-inch thicknesses. It can be obtained in various size sheets, in various densities, and with several treatments.

insulation, thermal—Any material high in resistance to heat transmission that, when placed in the walls, ceiling, or floors of a structure, will reduce the rate of heat flow.

interior finish—Material used to cover the interior framed areas or materials of walls and ceilings.

joint—The space between the adjacent surfaces of two members or components joined and held together by nails, glue, cement, mortar, or other means.

kiln-dried lumber—Lumber that has been kiln-dried often to a moisture content of 6 to 12 percent. Common varieties of softwood lumber, such as framing lumber, are dried to a somewhat higher moisture content.

knot—In lumber, the portion of a branch or limb

of a tree that appears on the edge or face of the piece.

lath—A building material of wood, metal, gypsum, or insulating board that is fastened to the frame of a building to act as a plaster base.

lumber—The product of the sawmill and planing mill not further manufactured other than by sawing, resawing, and passing lengthwise through a standard planing machine, crosscutting to length, and matching. Often graded (Fig. G-5).

lumber, boards—Yard lumber less than 2 inches thick and 2 or more inches wide.

lumber, dimension—Yard lumber from 2 inches to, but not including, 5 inch inches thick and 2 or more inches wide. Includes joists, rafters, studs, plank, and small timbers.

lumber, dressed size—The dimension of lumber after shrinking from green dimension and after machining to size or pattern.

lumber, matched—Lumber that is dressed and shaped on one edge in a grooved pattern and on the other in a tongued pattern.

lumber, shiplap—Lumber that is edge-dressed to make a close rabbeted or lapped joint.

lumber, timbers—Yard lumber 5 or more inches in least dimension. Includes beams, stringer, posts, caps, sills, girders, and purlins.

lumber, yard—Lumber of those grades, sizes, and patterns which are generally intended for ordinary construction, such as framework and rough coverage of houses.

Fig. G-5. Lumber grading (courtesy California Redwood Association).

mastic—A pasty material used as a cement or as a protective coating.

miter joint—The joint of two pieces at an angle that bisects the joining angle. For example, the miter joint at the side and head casing at a door opening is made at a 45-degree angle.

moisture content of wood—Weight of the water contained in the wood, usually expressed as a percentage of the weight of the ovendried wood.

molding—A wood strip having a curved or projected surface used for decorative purposes.

mortise—A slit cut into a board, plank, or timber, usually edgewise, to receive the tenon of another board, plank, or timber to form a joint.

natural finish—A transparent finish which does not seriously alter the original color or grain of the natural wood. Natural finishes are usually provided by sealers, oils, varnishes, water-repellent preservatives, and other similar materials.

nonbearing wall—A wall supporting no load other than its own weight (Fig. G-6).

on center (O.C.)—The measurement of spacing for studs, rafters, joists, and the like in a building from the center of one member to the center of the next.

panel—In house construction, a thin, flat piece of wood, plywood, or similar material framed by stiles and rails as in a door, or fitted into grooves of thicker material with molded edges for decorative wall treatment.

partition—A wall that subdivides spaces within a story of a building.

penny—As applied to nails, it originally indicated the price per hundred. The term now serves as a measure of nail length and is abbreviated by the letter *d*.

pith—The small, soft core at the original center of a tree around which wood formation takes place.

plywood—A piece of wood made of three or more layers of veneer joined with glue and usually laid

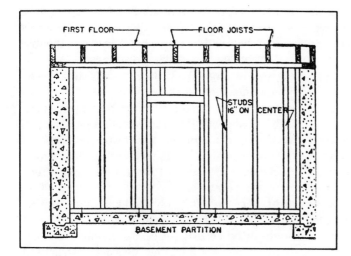

Fig. G-6. Nonload-bearing wall.

with the grain of adjoining plies at right angles. Almost always an odd number of plies are used to provide balanced construction.

preservative—Any substance that, for a reasonable length of time, will prevent the action of wood-destroying fungi, borers of various kinds, and similar destructive agents on wood.

primer—The first coat of paint in a paint job that consists of two or more coats; also the paint used for such a first coat (Fig. G-7).

putty—A type of cement usually made of whiting and boiled linseed oil, beaten or kneaded to the consistency of dough, and used in sealing glass in sash, filling small holes and crevices in wood, and for similar purposes.

quarter round—A small molding that has the cross section of a quarter circle.

rabbet—A rectangular longitudinal groove cut in the corner edge of a board or plank.

raw linseed oil—The crude product processed from flaxseed and usually without much subsequent treatment.

reflective insulation—Sheet material with one or both surfaces of comparatively low heat emissivity, such as aluminum foil. When used in building construction the surfaces face air spaces, reducing the radiation across the space.

sapwood—The outer zone of wood, next to the bark. In the living tree it contains some living cells (the heartwood contains none), as well as dead and dying cells. In most species it is lighter colored than the heartwood. In all species, it is lacking in decay resistance.

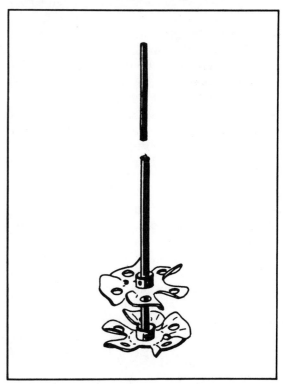

Fig. G-7. Paint mixer.

sealer—A finishing material, either clear or pigmented, that is usually applied directly over uncoated wood for the purpose of sealing the surface.

semigloss paint or enamel—A paint or enamel made with a slight insufficiency of nonvolatile vehicle so that its coating, when dry, has some luster but is not very glossy.

sheathing—The structural covering, usually wood boards or plywood, used over studs or rafters of a structure. Structural building board is normally used only as wall sheathing.

shellac—A transparent coating made by dissolving *lac,* a resinous secretion of the lac bug (a scale insect that thrives in tropical countries, especially India), in alcohol.

siding—The finish covering of the outside wall of a frame building, whether made of horizontal weatherboards, vertical boards with battens, shingles, or other material.

sill—The lowest member of the frame of a structure resting on the foundation and supporting the floor joists or the uprights of the wall. The member forming the lower side of an opening, such as a doorsill, windowsill, etc.

sleeper—Usually a wood member embedded in concrete, as in a floor, that serves to support and to fasten subfloor or flooring.

soffit—Usually the underside of an overhanging cornice.

solid lumber paneling—Decorative interior paneling made of solid pieces of lumber rather than plywood and installed on walls, ceilings, and cabinets.

span—The distance between structural supports such as walls, columns, piers, beams, girders, and trusses.

square—A unit of measure—100 feet square—usually applied to roofing material. Sidewall coverings are sometimes packed to cover 100 square feet and are sold on that basis.

stile—An upright framing member in a panel door.

stool—A flat molding fitted over the windowsill between jambs and contacting the bottom rail of the lower sash.

story—That part of a building between any floor and the floor or roof above it.

stringer—A timber or other support for cross members in floors or ceilings. In stairs, the support on which the stair treads rest.

strip flooring—Wood flooring consisting of narrow, matched strips.

stucco—Most commonly refers to an outside plaster made with Portland cement as its base.

stud—One of a series of slender wood or metal vertical structural members placed as supporting elements in walls and partitions. (Plural: studs or studding.)

subfloor—Boards or plywood laid on joists over which a finish floor is to be laid.

suspended ceiling—A ceiling system supported by hanging from the overhead structural framing.

termites—Insects which superficially resemble ants in size, general appearance, and habit of living in colonies; hence, they are frequently called "white ants." Subterranean termites establish themselves in buildings, not by being carried in with lumber, but by entering from ground nests after the building has been constructed. If unmolested, they eat out the woodwork, leaving a shell of sound wood to conceal their activities. Damage may proceed so far as to cause collapse of parts of a structure before discovery. There are about 56 species of termites known in the United States; but the two major ones, classified by the manner in which they attack wood, are ground-inhabiting or subterraneal termites (the most common) and dry-wood termites (found almost exclusively along the extreme southern border and the Gulf of Mexico in the United States).

termite shield—A shield, usually of noncorrodible metal, placed in or on a foundation wall or other mass of masonry or around pipes to prevent passage of termites.

toenailing—To drive a nail at a slant with the initial surface in order to permit it to penetrate into a second member.

tongue and groove (Fig. G-8)—See *dressed and matched.*

trim—The finish materials in a building, such as

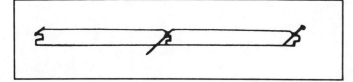

Fig. G-8. Tongue and groove.

moldings, applied around openings (window trim, door trim) or at the floor and ceiling of rooms (baseboard, cornice, and other moldings) (Fig. G-9).

truss—A frame or jointed structure designed to act as a beam of long span. Each member is usually subjected to longitudinal stress only, either tension or compression.

turpentine—A volatile oil used as a thinner in paints and as a solvent in varnishes. Chemically, it is a mixture of terpenes.

undercoat—A coating applied prior to the finishing or top coats of a paint job. It may be the first of two or the second of three coats. In some usage of the word, it may become synonymous with priming coat.

underlayment—A material placed under finish coverings, such as flooring, wall paneling, or

Fig. G-9. Trim (courtesy Hardwood Plywood Manufacturers Association).

shingles, to provide a smooth, even surface for applying the finish.

vapor barrier—Material used to retard the movement of water vapor into walls and prevent condensation in them. Usually considered as having a perm value of less than 1.0. Applied separately over the warm side of exposed walls or as a part of batt or blanket insulation.

varnish—A thickened preparation of drying oil or drying oil and resin suitable for spreading on surfaces to form continuous, transparent coatings, or for mixing with pigments to make enamels.

vehicle—The liquid portion of a finishing material; it consists of the binder (nonvolatile) and volatile thinners.

veneer—Thin sheets of wood made by rotary cutting or slicing of a log.

vermiculite—A mineral closely related to mica, with the faculty of expanding when heated to form lightweight material with insulation quality. Used as bulk insulation and also as aggregate in insulating and accoustical plaster and in insulating concrete floors.

volatile thinner—A liquid that evaporates readily and is used to thin or reduce the consistency of finishes without altering the relative volumes of pigments and nonvolatile vehicles.

wane—bark, or lack of wood from any cause, on the edge or corner of a piece of wood.

water-repellent preservative—A liquid designed to penetrate into wood and impart water repellency and a moderate preservative protection. It is used for millwork, such as sash and frames, and is usually applied by dipping.

wood rays—Strips of cells extending radially within a tree and varying in height from a few cells in some species to 4 inches or more in oak. The rays serve primarily to store food and to transport it horizontally in the tree.

Index

Other Bestsellers From TAB